Chongying Dong
James Lepowsky

Generalized Vertex Algebras and Relative Vertex Operators

Birkhäuser
Boston • Basel • Berlin

Chongying Dong
Department of Mathematics
University of California, Santa Cruz
Santa Cruz, CA 95064

James Lepowsky
Department of Mathematics
Rutgers University
New Brunswick, NJ 08903

Library of Congress Cataloging In-Publication Data

Dong, Chongying, 1958-
Generalized vertex algebras and relative vertex operators /
Chongying Dong, James Lepowsky.
p. cm.-- (Progress in mathematics ; v. 112)
Includes bibliographical references and index.
ISBN 0-8176-3721-4
1. Vertex operator algebras. I. Lepowsky, J. (James) II. Title.
III. Series: Progress in mathematics (Boston, Mass.) ; vol. 112.
QA326.D66 1993 93-21608
512'.55--dc20 CIP

Printed on acid-free paper

Birkhäuser

ISBN 0-8176-3721-4
ISBN 3-7643-3721-4
Typeset by the Authors in LATEX.
Printed and bound by Quinn-Woodbine, Woodbine, NJ.
Printed in the U.S.A.

9 8 7 6 5 4 3 2 1

Progress in Mathematics
Volume 112

Contents

Preface

In the past few years, vertex operator algebra theory has been growing both in intrinsic interest and in the scope of its interconnections with areas of mathematics and physics. The structure and representation theory of vertex operator algebras is deeply related to such subjects as monstrous moonshine, conformal field theory and braid group theory. Vertex operator algebras are the mathematical counterpart of chiral algebras in conformal field theory. In the Introduction which follows, we sketch some of the main themes in the historical development and mathematical and physical motivations of these ideas, and some of the current issues.

Given a vertex operator algebra, it is important to consider not only its modules (representations) but also intertwining operators among the modules. Matrix coefficients of compositions of these operators, corresponding to certain kinds of correlation functions in conformal field theory, lead naturally to braid group representations. In the special but important case when these braid group representations are one-dimensional, one can combine the modules and intertwining operators with the algebra to form a structure satisfying axioms fairly close to those for a vertex operator algebra. These are the structures which form the main theme of this monograph. Another treatment of similar structures has been given by Feingold, Frenkel and Ries (see the reference [FFR] in the Bibliography), and in fact the material developed in the present work has close connections with much work of other people, as we explain in the Introduction and throughout the text.

We present in this work a systematic treatment of several natural generalizations of the concept of vertex operator algebra. Detailed proofs are included, and many examples are interwoven through the axiomatic treatment, as motivation and as applications. We have tried to make the text quite accessible and self-contained, although the reader would probably find that some exposure to *Vertex Operator Algebras and the Monster* by Frenkel, Lepowsky and Meurman (see [FLM3]) and to *On Axiomatic Approaches to Vertex Operator Algebras and Modules* by Frenkel, Huang and Lepowsky (see [FHL]) would be helpful. In fact, many of the ideas discussed here are natural continuations of ideas presented in those works. This monograph was initially intended to be a paper supplying the details of our proof of a "generalized Jacobi identity" for "relative vertex operators," which we announced [DL1] in the course of explaining the essential equivalence between the mathematical notion of "Z-algebra" and the physical notion of "parafermion algebra." But in the process of expressing our proof in a natural axiomatic context, we decided in addition to discuss systematically how a variety of structures are illuminated by similar or closely related considerations, and our research paper gradually expanded into the present work. We call the most general of our axiomatic structures "abelian intertwining algebras." Their theory, as presented here, includes the theory of vertex operator algebras themselves as a special case, and in fact our development serves to clarify some important aspects of that theory.

Versions of this work have been used as the basis of some graduate courses and seminars at Rutgers University over the past few years, and our experience has suggested that it would be more useful to publish it as a single monograph than as a series of individual papers. Any combination of the following chapters could serve as a focus of a course or seminar, with the preceding material in this work invoked as necessary: Chapters 6, 7 and 8, which deal with "generalized vertex operator algebras" and their close relations with the operator product expansion in conformal field theory and with one-dimensional monodromy representations of braid groups; Chapter 9, which presents the analogous more general theory for "generalized vertex algebras"; Chapter 12, whose subject is the analogous treatment of the still more general concept of abelian intertwining algebra, including a study of certain algebras of intertwining operators for vertex operator algebras and also of certain very closely related third cohomology groups; Chapter 13, devoted to the vertex-operator-algebraic structure underlying certain modules for affine Lie algebras; or Chapter 14, which concludes the monograph with the development and synthesis of the concepts of Z-algebra and parafermion algebra, based on the axiomatic theory of generalized vertex operator algebras. The Introduction and the individual chapter introductions provide guides to all these topics, including motivation, background and discussions of related work.

Acknowledgments. We would like to thank the people who attended the Rutgers lectures for their enthusiasm and for pointing out misprints, and we particularly thank Haisheng Li for simplifying an argument in Chapter 13. We are also grateful to Charles Weibel for comments on group cohomology and especially to Myles Tierney for drawing our attention to the Eilenberg-Mac Lane cohomology theory of abelian groups, which plays an important role in Chapter 12. We gratefully acknowledge the following partial support during the several years in which this work was carried out: C.D.: a postdoctoral fellowship at Rutgers University and grants from the National Science Foundation and the Regents of the University of California; J.L.: a Guggenheim Foundation Fellowship, the Rutgers University Faculty Academic Study Program and grants from the National Science Foundation; J.L. also thanks the Institute for Advanced Study and the Mathematical Sciences Research Institute for their hospitality.

Finally, we owe special thanks to Ann Kostant for arranging the publication of this monograph with Birkhäuser and for making the final preparation for publication a far more pleasant task than we imagined it could possibly be.

Chongying Dong and James Lepowsky

Chapter 1

Introduction

The rapidly-evolving theory of vertex operator algebras provides deep insight into many important algebraic structures, including the Monster finite simple group and highest weight modules for affine Lie algebras and for the Virasoro algebra. The original motivation for the introduction of the precise notion of vertex operator algebra arose from the problem of realizing the Monster as a symmetry group of a natural infinite-dimensional structure, and in the book [FLM3], the Monster was in fact realized as the automorphism group of a certain vertex operator algebra. Partly motivated by the announcement [FLM1], Borcherds introduced a notion of "vertex algebra" in his announcement [Bor], and the variant of this notion that we call "vertex operator algebra," equipped with the fundamental "Jacobi identity" as the main axiom, was introduced in [FLM3] and [FHL]. (The reader may wish to consult the introductory material in these works for historical background.) Vertex operator algebras can be viewed as "complex analogues" of both Lie algebras and associative algebras. In the physics literature, a physical counterpart of the notion of vertex operator algebra — the notion of what is now usually called "chiral algebra" — evolved in relation to dual resonance theory and more recently, to string theory and conformal field theory; see for instance for the historical comments in [FLM3] and [MS]. In fact, much of conformal field theory, basic algebraic features of whose axiomatic structure were formalized in such works as [BPZ] and later, [MS], can be understood in terms of the theory of vertex operator algebras and their representations (or modules). The representation theory of vertex operator algebras, via the monodromy and "half-monodromy" associated with correlation functions, which are essentially matrix coefficients of products of intertwining operators among modules, provides a natural context for studying the kinds of braid group representations that enter into the construction of topological invariants like the Jones polynomial; this was made clear as a result of such investigations as [TK] and [MS].

In the present monograph, we generalize the theory of vertex operator algebras and their modules in a systematic way so as to incorporate one-dimensional braid group representations intrinsically into the algebraic structure. Our original motivation was to place the theory of "Z-algebras" into an elegant axiomatic context, and in particular, to embed such algebras into larger, more natural algebras; Z-algebras have played a central role in the vertex-operator-theoretic interpretation of generalized Rogers-Ramanujan partition identities (see below). But it turns out that the theory presented here also enables us to treat a range of interesting subjects from a unified viewpoint, including also the theory of vertex operators associated with rational lattices; the theory of intertwining operators for vertex operator algebras based on even lattices and, to a certain extent, for those based on standard (irreducible) modules (that is, integrable highest weight modules) for affine Lie algebras; and third cohomology in the Eilenberg-Mac Lane "cohomology theory of abelian groups." (This cohomology theory, which until now has not been commonly used, is different from ordinary group cohomology theory.) The families of "generalized vertex operator algebras" that we introduce in order to enrich the theory of Z-algebras and develop its appropriate foundations are based on the construction of what we call "relative (untwisted) vertex operators" and on our "generalized Jacobi identity" for these operators, which we announced in [DL1]. In addition to supplying the details of the results announced in [DL1], this work considerably expands the scope of those results by developing them axiomatically; the generalized Jacobi identity is the main axiom for generalized vertex operator algebras.

Structures similar to generalized vertex operator algebras, and to what we call "generalized vertex algebras" below, have been independently introduced and studied in [FFR], with different motivations, examples (involving spinor constructions) and axiomatic approach from ours, and also in [Mo]. The notion of generalized vertex algebra includes as very special cases various notions of ($\mathbf{Z}/2\mathbf{Z}$-graded) "vertex operator superalgebra," versions of which are discussed in [FFR], [Go] and [T], for example. More generally, it includes other useful variants of the notion of vertex operator algebra for which the associated monodromy is trivial, such as the $\mathbf{Z}$-graded superalgebraic structures considered in such works as [LZ] and the "colored vertex operator superalgebras" considered in [X]. In particular, the axiomatic treatment in the present work applies to these structures.

Actually, three distinct levels of generalization of the concept of vertex operator algebra (and modules) are presented in this monograph, as we explain below; the flexibility in the choice of axiom system illustrates some of the considerable richness of the subject. The main results and examples presented here concerning the most general of these concepts — that of what we call "abelian intertwining algebra" — have been announced in [DL2], where we have introduced this notion and described in a self-contained way the connection between vertex operator algebras and the cohomology theory of abelian groups. In a sequel [DL3] to the present work, we present a general theory of relative *twisted* vertex operators, based on lattice automorphisms, generalizing and incorporating the results of [L2] and [FLM2].

The concept of vertex operator algebra can be thought of as a solution to the following (very roughly formulated) general problem: Find a natural algebraic structure which implements the operator equation

$$A(z_1)B(z_2) = B(z_2)A(z_1), \tag{1.1}$$

where A and B are two operators parametrized by complex numbers z_1 and z_2, respectively, and the equation is understood in the sense of analytic continuation; we think of the left-hand side as being defined only in the domain $|z_1| > |z_2|$ and the right-hand side only in the opposite domain $|z_2| > |z_1|$. This coarsely-expressed idea is in fact realized in a precise way in the "duality" theory of vertex operator algebras, in particular, in the notion of what is called "commutativity" of the "vertex operators" $A(z_1)$ and $B(z_2)$. One also considers an analogous "associativity" relation, which amounts to the "associativity of the operator product expansion" in conformal field theory. Such commutativity and associativity relations lead to commutator relations for the operators which are the Laurent series coefficients of the vertex operators $A(z_1)$ and $B(z_2)$, and in fact to a family of more general relations which can all be combined in compact generating-function form into a single "Jacobi(-Cauchy) identity." This Jacobi identity for vertex operator algebras, which concentrates a great deal of information in one formula, was so named (in [FLM3]) because of its analogy with the Jacobi identity for Lie algebras. Briefly, a vertex operator algebra is an infinite-dimensional vector space acted on by "vertex operators," which are parametrized by both the elements of the vector space and by a formal variable, subject to a number of axioms, the most important of which is the Jacobi identity. The vertex operators provide the "multiplication," although a vertex operator algebra is not an algebra, even a nonassociative algebra, in the ordinary sense. "Duality" is a conformal-field-theoretic term with roots in the old dual resonance model.

The systematic theory of vertex operator algebras and modules, based on the Jacobi identity and including the relations between this identity and the more intrinsically geometric commutativity and associativity relations, is presented in [FLM3] and [FHL]. In the course of generalizing this theory in the present work, we shall review in a reasonably self-contained way the important basic features of the theory of vertex operator algebras themselves, including examples, so that the reader of this monograph will not need more than a bit of familiarity with some of the more elementary material in [FLM3] and [FHL]. All the fundamental concepts such as the Jacobi identity and vertex-operator commutativity and associativity will be explained fully in the body of this work. Our underlying setup is the formal-variable calculus developed in [FLM3] and [FHL] (and recalled in this work). Vertex operators and their properties, such as the Jacobi identity, are expressed in terms of formal Laurent series. Rational functions and algebraic functions of complex variables also play crucial roles, after the formal-calculus foundations have been laid.

We shall now explain something about our first motivation for generalizing this theory: There are important algebraic structures — Z-algebras and "algebras of parafermion fields" — which admit the more general operator equation

$$A(z_1)B(z_2) = e^{i\pi r}B(z_2)A(z_1), \tag{1.2}$$

where r is a rational number depending on A and B, and the "phase" $e^{i\pi r}$ is to arise as a "half-monodromy" factor; when the operators $A(z_1)$ and $B(z_2)$ are permuted a second time, the result is the monodromy factor $e^{2\pi ir}$. Z-algebras (the symbol "Z" referring to the fact that these algebraic structures centralize the actions of certain Heisenberg subalgebras of affine Lie algebras) were originally introduced and studied in [LW2]-[LW5] and [LP1]-[LP3], before conformal field theory became fashionable in physics through the influence of such works as [BPZ], to study the representation theory, including the theory of vertex operator representations, of arbitrary affine Kac-Moody Lie algebras. In fact, they were introduced in the course of a program of constructing bases of standard modules for these algebras, in connection with supplying vertex-operator-theoretic interpretations of combinatorial identities such as the Rogers-Ramanujan identities and generalizations due to Gordon, Andrews and Bressoud. This program, as first carried out in a family of cases in [LW2], has had three main stages: constructing "Z-operators" in terms of ordinary (possibly twisted) vertex operators, so as to centralize an underlying Heisenberg algebra; establishing "generalized commutator and anticommutator relations" for these operators; and (typically the hardest part) using such relations to construct bases (compatible with a natural Z-algebra filtration) of the vacuum spaces for certain Heisenberg algebras in the irreducible modules under consideration, thereby implementing the combinatorial identities in terms of character formulas for irreducible modules. These bases led for example, in [LP2], to new formulas involving "string functions" for $A_1^{(1)}$ (such functions having been originally introduced, and computed for level-one and some higher-level standard modules, in [FL]). We remark that for certain steps in this program, the higher-level standard modules were embedded either into tensor products of level-one standard modules or into level-one standard modules themselves, by restricting to a suitable subalgebra of the affine Lie algebra, but the main goal was the much more subtle one of finding precise bases for the *irreducible* higher-level modules. In the present work, the first two of the three main stages of this program are placed into a natural, general context. Progress in the original Z-algebra program, including that made in [C], [Hu], [Ma], [MP], [Mi1]-[Mi4] and [P1]-[P2], is reviewed in [L3].

The main feature of Z-algebras is that the basic algebraic relations are "generalized commutator relations" for $A(z_1) = \sum A_n z_1^n$ and $B(z_2) = \sum B_n z_2^n$, in place of Lie algebra commutator relations. Such relations are of the following typical shape:

$$(1.3)\quad (z_1 - z_2)^r A(z_1)B(z_2) - (z_2 - z_1)^r B(z_2)A(z_1) = C(z_1, z_2),$$

where the variables z_1 and z_2 are now to be understood as commuting formal variables rather than complex variables, r is a rational number as above, $(z_1 - z_2)^r$ and $(z_2 - z_1)^r$ are understood as the binomial expansions in nonnegative integral powers of z_2 and z_1, respectively, and $C(z_1, z_2)$ is an operator-valued formal Laurent series (cf. (1.2)). Equating the coefficients of the monomials in z_1 and z_2 in the Z-algebra relation (1.3) gives a family of generalized commutator relations in component form, each relation expressing an infinite linear combination of products of the operators A_n and B_n as a certain single operator. Families of relations of this type were used in such works as [LW2]-[LW5] to "straighten" monomials in the coefficient operators such as A_n, leading to a crucial difference-two condition for the subscripts n occurring in the resulting straightened monomials. An analogous difference-one condition would give us a "Pauli exclusion principle" for the corresponding monomials, and in fact this actually occurs in certain easy cases, such as that of the level-two standard $A_1^{(1)}$-modules. Instead, however, there is generically a difference-*two* condition, compatible, in fact, with the Rogers-Ramanujan identities; this condition is in effect a "generalized Pauli exclusion principle." Its first real occurrence is for level-three standard $A_1^{(1)}$-modules. The summands in the Rogers-Ramanujan identities and their generalizations correspond exactly to natural subquotients in the Z-algebra filtration.

Equations of the type (1.1) arise naturally in conformal field theory and statistical mechanics (see e.g. [FradK]). In fact, the same algebraic structure discussed above also arose in the "parafermionic conformal field theory" of [ZF1], in the form of "nonlocal current algebras"; the operators satisfying these relations in [ZF1] are the same as the corresponding (untwisted) operators in [LP1], as we explain in detail in this work. A "dictionary" between the mathematical and physical settings was first provided in [DL1], where it was in fact announced that parafermion algebras are essentially equivalent to Z-algebras in a precise sense; such a correspondence was already hinted in [ZF1]. Different aspects and applications of this common theory were developed in the two different contexts; the works on Z-algebras used the generalized commutator relations and analogous types of identities to construct bases of the spaces on which the operators act irreducibly, in connection with the interpretation and proof of partition-function identities, as we have mentioned, while physics papers on parafermion algebras focused on the conformal-field-theoretic structure, including the role of the Virasoro algebra. Incidentally, the term "parafermion" essentially comes from the relations of the type (1.2) and a related "fractional-spin" property, but the difference-two conditions and generalized Pauli exclusion principle of say [LW2], referred to above, in retrospect serve to amplify the significance of this term. The conformal-field-theoretic structure studied in [ZF2] corresponds precisely to a case of the theory of relative *twisted* vertex operators (cf. [LW1]-[LW5], [Hu], [DL3]). The Z-algebra theory for *arbitrary* affine Lie algebras introduced and studied in [LW2]-[LW5] and [LP1]-[LP3] was used in the conformal-field-theoretic setting in [Gep].

The phase $e^{i\pi r}$ being nontrivial in general, the theory of vertex operator algebras does not directly apply to these algebraic structures. In this monograph, we present systematic generalizations of the notion of vertex operator algebra which include Z-algebra-type structure as natural generating substructure and which incorporate the (half-) monodromy associated with one-dimensional braid group representations. Just as the commutator relations for vertex operators are embedded into a much more natural and general Jacobi identity in the theory of vertex operator algebras, we embed the generalized commutator relations for Z-algebras into a much larger algebraic structure satisfying a powerful "generalized Jacobi identity," which serves as the main axiom for what we call "generalized vertex operator algebras" (Chapter 6). We also introduce two successively more general notions — those of "generalized vertex algebra" (Chapter 9) and of "abelian intertwining algebra" (Chapter 12), each based on an appropriate generalized Jacobi identity, and each designed to handle families of important and interesting examples. We prove that the respective Jacobi identities are equivalent, in the presence of more elementary axioms, to "generalized commutativity" relations of the type (1.2). In this spirit, we see that the very notion of abelian intertwining algebra is in some sense naturally forced on us when we propose to consider a system of generalized commutativity relations. We also systematically discuss the role of analogous "generalized associativity" relations. The axioms for a generalized vertex operator algebra appear fairly close to those for a vertex operator algebra. While the axioms for the other two structures look somewhat more complicated, the theory is quite similar, and the greater generality is really needed to study certain very natural situations. On the other hand, we decided that it would be useful to begin with a relatively simple-looking set of axioms, and to build toward greater generality as the need became apparent. Each variant of the notion of generalized vertex operator algebra carries associated natural notions of module, and we study these as well.

We conclude this Introduction with a brief description of the contents of this work, chapter by chapter, including comments on applications of the various levels of generality. The reader is also urged to consult the introductions to the individual chapters for further details, motivation and historical remarks.

In Chapter 2, we present a basic setting, involving a (rational) lattice and the (finite-dimensional complex) vector space which it generates. One important ingredient is a subspace of the vector space with respect to which the vertex operators based on a certain infinite-dimensional vector space will be "relativized." In Chapter 3, we introduce the notion of relative untwisted vertex operators, both parametrized by this infinite-dimensional vector space and acting on it, and we discuss the elementary properties of these operators. The "unrelativized" degenerate case provides a detailed review of the usual vertex operators associated with a lattice. "Quotient vertex operators" with respect to certain abelian groups are constructed in Chapter 4. These operators will provide the underlying setup for the basic examples of generalized vertex operator algebras. While some of the notation in the first few chapters might seem technically complicated, it turns out that the generality is very natural; also, in the individual chapter introductions, we tell those readers interested in only certain applications how to safely ignore certain details of the setup.

Chapter 5, which is central to this work, focuses on the formulation and proof of our generalized Jacobi identity for relative vertex operators. The main difference between this identity and the various forms of the Jacobi identity in [FLM3] which this identity generalizes is that this identity involves certain algebraic functions, whose source is the nonintegrality of the "inner products" of lattice elements. We present several useful consequences of the main Jacobi identity, by passing to special cases and to the quotient structure of Chapter 4.

In Chapter 6 we present the definitions of the notions of generalized vertex operator algebras and their modules, and we give a number of elementary properties. These structures are infinite-dimensional vector spaces graded both by the rational numbers and by a finite abelian group G equipped with a suitable symmetric bilinear form, and their Jacobi identity includes algebraic functions built from this form. The subspace corresponding to the identity element of G is essentially a vertex operator algebra. We summarize most of the main points in the earlier chapters by formulating theorems asserting that the structures already developed, based on relative vertex operators, provide examples of generalized vertex operator algebras and of modules. We also explain the precise sense in which the notion of (ordinary) vertex operator algebra, in the terminology of [FLM3] or [FHL], is a special case. In addition, we present certain reasonable generalizations of the notion of module, in particular, a generalization based on G-sets, which will be very useful in later chapters in the study of intertwining operators, and a generalization allowing a $\mathbf{C}$-grading rather than a $\mathbf{Q}$-grading; the restriction to $\mathbf{Q}$-grading is usually not important. Finally, we give a definition of "vertex operator superalgebra" as a special case of the notion of generalized vertex operator algebra with $G = \mathbf{Z}/2\mathbf{Z}$, and we point out that certain generalized vertex operator algebras based on relative vertex operators provide examples of vertex operator superalgebras.

In Chapter 7 we formulate and discuss the fundamental "duality" properties of generalized vertex operators algebras and modules, namely, generalized rationality, commutativity and associativity. These important properties, which provide a link between the Jacobi identity and more geometric concepts, essentially assert that suitable "matrix coefficients" of products and iterates of vertex operators, after multiplication by the formal series expansions of certain algebraic functions, become the formal series expansions of common rational functions in different domains. These duality properties generalize the corresponding properties for ordinary vertex operator algebras, for which the algebraic functions become trivial. Formulas (1.1) and (1.2) above are roughly-expressed versions of such commutativity relations for vertex operator algebras and generalized vertex operator algebras, respectively. We also show that the duality properties imply the Jacobi identity in the presence of more elementary assumptions, and that these properties can therefore be taken as alternatives to the Jacobi identity in the definition of generalized vertex operator algebra or module. Moreover, in the case of algebras, we establish the important fact that the generalized commutativity relation alone is sufficient to imply the associativity relation and hence the Jacobi identity. For both algebras and modules, we provide a link between the Jacobi identity and the duality relations, by means of "generalized commutator and associator relations," of which formula (1.3) above is a roughly-expressed prototype. Generalized commutativity is also extended to products of more than two vertex operators. It is important to note that when applied to the special case of (ordinary) vertex operator algebras, the treatment of both the formulations and the proofs of duality results in this chapter offers some useful refinements and simplifications of the corresponding treatment of such duality in [FLM3] and [FHL]. For instance, we can now show directly that (generalized) rationality and commutativity for pairs of vertex operators imply the same for several operators. This is accomplished by the use of natural conditions on the orders of certain poles, in the statements of the generalized rationality conditions, and the use of natural reformulations of such conditions as the generalized commutativity condition in a form not involving "matrix coefficients" (see especially Proposition 7.6, formula (7.24) and Remark 7.7). We should point out here that in particular, for generalized vertex operator algebras, the Jacobi identity can (in the presence of simpler assumptions) be replaced by the simple version of the generalized commutator relation given in formula (7.24). (The analogous comment holds just as well for our other two axiomatic structures — generalized vertex algebras and abelian intertwining algebras.) Applying this result to the important case of (ordinary) vertex operator algebras immediately gives us the result that the (ordinary) Jacobi identity can be replaced by the simple relation (which is just (7.24) for the case of vertex operator algebras)

$$(1.4)\qquad (z_1 - z_2)^n [Y(v_1, z_1), Y(v_2, z_2)] = 0,$$

where z_1 and z_2 are commuting formal variables, v_1 and v_2 are arbitrary elements of the algebra, $Y(v_i, z_i)$ are the corresponding vertex operators and n is a sufficiently large positive integer depending on v_1 and v_2. This result (and formula (7.24)) turn out to be very useful. For instance, formula (7.24) is used in [Gu] to give a direct proof of the Jacobi identity for tensor product algebras and modules, and also in [A] to prove a certain Jacobi identity. Our treatment also includes analogues of formula (1.4) for "(generalized) associators" in place of commutators.

Chapter 8 is devoted to showing how the monodromy associated with the products of several vertex operators, through the monodromy properties of certain algebraic functions, naturally produces one-dimensional braid group representations, when the formal variables are specialized to nonzero complex variables. These representations factor through certain Hecke algebras. The rational number r in formula (1.2) is the prototype of a parameter defining the braid-group representations.

In Chapter 9 we consider a more general algebraic structure, "generalized vertex algebras," which include the vertex-operator-algebraic structure arising from a rational lattice (not necessarily positive definite) as well as the generalized vertex operator algebras considered in Chapter 6, and in addition, a variant of what Borcherds calls "vertex algebras" [Bor]. The main differences between the axioms for generalized vertex algebras and the axioms for the structures studied in Chapter 6 is that two grading-restriction axioms are removed, the Jacobi identity acquires a numerical factor associated with an alternating bilinear form on the grading group G, and G need not be finite. The duality and monodromy results in Chapters 7 and 8 remain valid, with suitable modifications, for generalized vertex algebras. Using the result that the Jacobi identity follows from generalized commutativity, we sketch an alternative proof of the main Jacobi identity of Chapter 5, and hence of the important special cases of this result, Theorems 8.6.1 and 8.8.23 of [FLM3]. (We remark that the right-hand side of the Jacobi identity given in Theorem 8.8.23 of [FLM3] and in [DL1] contains a misprint, corrected in the present work; in the proof and in all applications of that identity in [FLM3], it was in fact the correct formula which was actually proved and which was actually used.) We observe that essentially the entire discussion of duality applies, with obvious modifications, to modules. Finally, we define a notion of "vertex superalgebra," which includes the notion of vertex operator superalgebra considered in Chapter 6, as a special case of generalized vertex algebra with $G = \mathbf{Z}/2\mathbf{Z}$.

In the brief Chapter 10, we define the notion of tensor product of generalized vertex algebras, using the duality results to prove that the tensor product structure is in fact a generalized vertex algebra. We also introduce the notion of tensor product module for such a tensor product algebra. These constructions generalize the corresponding ones given in [FHL].

In Chapter 11, we define the notion of intertwining operators among three modules for a generalized vertex algebra and related notions. We also establish generalized rationality, commutativity and associativity for pairs of operators of which one is a vertex operator parametrized by the given algebra and the second is an intertwining operator relating three given modules. As an application of such duality, we prove that in a setting involving irreducible modules, an intertwining operator is uniquely determined by the restriction of the operator to a pair of nonzero vectors. This result is useful even in the case of ordinary vertex operator algebras. Intertwining operators for (ordinary) vertex operator algebras were defined in [FHL].

Chapter 12 is devoted to the third of our three levels of generality – that of abelian intertwining algebra. We motivate this by first discussing the algebraic structure of the vertex algebra formed from an even lattice and a family of its modules together with a family of intertwining operators, which we construct by appropriately modifying the "obvious" vertex operators. By establishing a Jacobi identity for these intertwining operators, we find that we have a structure similar to a generalized vertex algebra, but incorporating a certain 3-cocycle (in the sense of ordinary group cohomology theory) on the abelian group labeling the irreducible modules. Motivated by this structure, we then present our formal definition of abelian intertwining algebra, and we formulate a precise theorem stating that the intertwining operator structure provides an example of an abelian intertwining algebra. The axiom system is intimately related to third cohomology in the Eilenberg-Mac Lane cohomology theory of abelian groups, and in fact abelian intertwining algebras provide natural operator realizations of such cohomology. Besides an ordinary 3-cocyle, an abelian 3-cohomology class also incorporates "braiding data." Together, this information is essentially equivalent to the (one-by-one) braiding and fusing matices which arise in families of conformal field theories based on abelian groups, in the sense explained in [MS], and in fact abelian intertwining algebras implement such conformal field theories in a precise way. We present the basic properties of abelian intertwining algebras, including the important duality and monodromy properties and the existence of tensor products of algebras and modules. These properties provide a kind of summary of all the general, axiomatic considerations in this entire work. An abelian intertwining algebra is essentially made up of the direct sum of a vertex algebra and a suitable family of modules for it, and the "multiplication structure" is given by intertwining operators satisfying generalized commutativity relations like (1.2). We show how the canonical quadratic form determined by (and determining) the abelian 3-cohomology class describes the natural one-dimensional braid group representations, which are also Hecke-algebra representations, arising in the usual way. For the motivating example based on an even lattice, we also relate the 3-cohomology class to certain "crossed modules," in the spirit of group cohomology theory. We also point out that the abelian intertwining algebras constructed from lattices provide an important family of examples of vertex superalgebras – those based canonically on integral lattices; these are analogues of the vertex algebras based canonically on even lattices ([Bor], [FLM3]).

In Chapter 13 we study the vertex operator algebras and their modules arising naturally from the higher-level standard representations of affine Lie algebras of types A, D and E. We first consider the untwisted space formed form the direct sum of l copies of the weight lattice of the underlying simple Lie algebra, and we thus focus our attention on the level l standard modules which can be embedded as irreducible components of tensor products of l basic modules. We establish that a distinguished one of these modules can be naturally endowed with the structure of a vertex operator algebra. The Virasoro algebra corresponds canonically to the Casimir element associated with the simple

Lie algebra, as we show by means of a useful general principle which, in the context of an arbitrary vertex operator algebra and module, indicates (using the Jacobi identity) that the vertex operator associated with a symmetrized product of two weight-one elements can be expressed as a normal ordered product of two vertex operators, each generated by one of the weight-one vectors. The normal-ordered quadratic nature of the Virasoro algebra operators, and also their bracket relations, are direct consequences. Next we consider the decomposition of the untwisted space as a direct sum of standard modules of level l, and we show that after we perform a suitable grading-shift, each of these standard modules acquires the structure of an irreducible module for the vertex operator algebra. We show conversely, using another useful general principle, concerning powers of a vertex operator whose components mutually commute, that every irreducible module for the vertex operator algebra carries the structure of a standard module of level l. Next we explicitly construct a family of canonical intertwining operators (in the precise sense of [FHL]) among the modules, by using the abelian-intertwining-algebra structure developed in Chapter 12, and by further modifying the resulting vertex operators so as to account properly for the grading. This construction directly exhibits intertwining operators satisfying the relevant Jacobi identity as well as the conditions satisfied by the "(chiral) vertex operators" proved to exist in [TK] (cf. [KZ] and [MS]), and in fact our method provides a natural way to construct explicit intertwining operators for a variety of conformal field theories. A number of results in this chapter overlap with some results in [FZ], although the methods are different, and the whole picture is a precise algebraic formulation of many of the fundamental features of the Wess-Zumino-Novikov-Witten model in conformal field theory.

Chapter 14, which combines the main results of the early chapters with those of Chapter 13, is devoted to the original goal of placing Z-algebras and parafermion algebras into an elegant axiomatic context based on relative vertex operators. We begin with the setup of Chapter 13, which we relativize to the Cartan subalgebra, and we correspondingly focus our attention on the vacuum spaces of standard modules with respect to a certain natural Heisenberg algebra. We show that certain quotient spaces of the vacuum spaces of the distinguished level l modules carry generalized vertex operator algebra structure, equipped with the quotient relative vertex operators introduced in Chapter 4, and that the corresponding quotient spaces of the vacuum spaces of general standard modules form modules for these generalized vertex operator algebras. We illustrate the theory by specializing our structure to the case of the affine Lie algebra $A_1^{(1)}$, and in this case we explain in considerable detail the essential equivalence between Z-algebras ([LP1]-[LP3]) and parafermion algebras [ZF1] by realizing the parafermion algebras as canonically modified Z-algebras acting on certain quotient spaces of vacuum spaces of standard modules, a construction carried out in [LP1]. This provides a precise mathematical foundation for "parafermion conformal field theory." Our finite abelian grading group G becomes the cyclic symmetry group in [ZF1]. The concluding results are a system of precise formulas of the the type (1.3), extracted as very special cases of a generalized Jacobi identity, recovering certain basic relations in [ZF1].

Chapter 2

The setting

In this chapter, we establish some basic assumptions which will remain in force essentially throughout this monograph. The main pieces of data are: a lattice L; an isometry ν of L; a central extension $\hat{L}$ of L by a finite cyclic group, determined by its associated commutator map, which is an alternating bilinear form on L; and a subspace $\mathbf{h}_*$ of the complex span $\mathbf{h}$ of L. The lattice and its central extension will be used to build the vector space V_L which will parametrize the basic vertex operators and on which these operators will also act. The isometry will not play a large role in this monograph; it will be important in the sequel [DL3] on relative twisted vertex operators. Considering only the trivial case $\nu = 1$ will be adequate for many purposes and will simplify some notation throughout this work. The subspace $\mathbf{h}_*$ is the subspace with respect to which we shall "relativize" the usual untwisted vertex operators. The "unrelativized" case, in which $\mathbf{h}_* = 0$, is still important, and the reader will find that specializing to this case will simplify quite a bit of the notation throughout this work, while many of the results will remain significant.

The symbols $\mathbf{Z}$, $\mathbf{N}$, $\mathbf{Q}$, $\mathbf{R}$ and $\mathbf{C}$ will be used to denote the integers, nonnegative integers, rational numbers, real numbers and complex numbers, respectively.

We work in the setting of [L2] and [FLM3]. Vector spaces will be over $\mathbf{C}$ unless another field is specified. We introduce the following notation: For each positive integer n, consider the standard primitive n^{th} root of unity

(2.1) $$\kappa_n = e^{\frac{2\pi i}{n}} \in \mathbf{C}^\times.$$

Using the general notation $\langle \cdot \rangle$ for "group generated by," we have the n-element group $\langle \kappa_n \rangle$. We fix a positive integer q and we assume the following data and conditions:

2.1. Let L be a (rational) lattice, i.e., a finite-rank free abelian group equipped with a symmetric $\mathbf{Q}$-valued $\mathbf{Z}$-bilinear form $\langle \cdot, \cdot \rangle$, not necessarily positive definite. We also assume that $\langle \cdot, \cdot \rangle$ is nondegenerate.

2.2. Let ν be an isometry of L such that $\nu^q = 1$ (but q need not be the exact order of ν and we may take $\nu = 1$).

2.3. Let c_0 be a $\mathbf{Z}$-bilinear map

$$c_0 : L \times L \to \mathbf{Z}/q\mathbf{Z}, \tag{2.2}$$

which is alternating, i.e.,

$$c_0(\alpha, \alpha) = 0 \tag{2.3}$$

for all $\alpha \in L$; then

$$c_0(\alpha, \beta) = -c_0(\beta, \alpha) \tag{2.4}$$

for all $\alpha, \beta \in L$. We also assume that

$$c_0(\nu\alpha, \nu\beta) = c_0(\alpha, \beta). \tag{2.5}$$

Remark 2.1: Viewed as an alternating $\mathbf{Z}$-bilinear map from $L \times L$ to $\mathbf{Z}/q\mathbf{Z}$, c_0 determines a central extension of L,

$$1 \to \langle \kappa_q \rangle \to \hat{L} \xrightarrow{\bar{\ }} L \to 1, \tag{2.6}$$

uniquely up to equivalence, by the commutator condition

$$aba^{-1}b^{-1} = \kappa_q^{c_0(\bar{a}, \bar{b})} \in \langle \kappa_q \rangle \tag{2.7}$$

for all $a, b \in \hat{L}$; we call c_0 the *commutator map* of the extension $\hat{L}$ (cf. [FLM3], Section 5.2). Moreover, condition (2.5) implies that ν lifts to an automorphism $\hat{\nu}$ of $\hat{L}$ fixing κ_q, that is,

$$\overline{\hat{\nu}a} = \nu\bar{a} \quad \text{for } a \in \hat{L}, \quad \hat{\nu}\kappa_q = \kappa_q, \tag{2.8}$$

and such a lifting is unique up to multiplication by a lifting of the identity automorphism of L, which is of the form

$$\begin{aligned} \lambda^* : \hat{L} &\to \hat{L} \\ a &\mapsto a\kappa_q^{\lambda(\bar{a})} \end{aligned} \tag{2.9}$$

for some $\lambda \in \mathrm{Hom}(L, \mathbf{Z}/q\mathbf{Z})$ (cf. [FLM3], Section 5.4). If $\nu = 1$, then we may take $\hat{\nu} = 1$.

Remark 2.2: If L is even (i.e., $\langle\alpha,\alpha\rangle \in 2\mathbf{Z}$ for all $\alpha \in L$) and if q is even, then L always has a central extension by $\langle\kappa_q\rangle \supset \langle\kappa_2\rangle = \langle -1\rangle$ such that

$$c_0(\alpha,\beta) = \langle\alpha,\beta\rangle + 2\mathbf{Z}. \tag{2.10}$$

2.4. Let $\mathbf{h}_*$ be a subspace (possibly 0 or $\mathbf{h}$) of

$$\mathbf{h} = \mathbf{C} \otimes_{\mathbf{Z}} L \tag{2.11}$$

on which the natural (nonsingular) extension of the form $\langle\cdot,\cdot\rangle$ on L, still denoted $\langle\cdot,\cdot\rangle$, remains nonsingular. That is,

$$\mathbf{h} = \mathbf{h}_* \oplus \mathbf{h}_*^{\perp}, \tag{2.12}$$

$^{\perp}$ denoting orthogonal complement. We write

$$\begin{array}{ll} \mathbf{h} \to \mathbf{h}_*^{\perp} & \quad \mathbf{h} \to \mathbf{h}_* \\ h \mapsto h', & \quad h \mapsto h'' \end{array} \tag{2.13}$$

for the projection maps to $\mathbf{h}_*^{\perp}$ and $\mathbf{h}_*$. We also assume that $\mathbf{h}_*$ is stable under the natural action of ν on $\mathbf{h}$:

$$\nu\mathbf{h}_* = \mathbf{h}_*. \tag{2.14}$$

Then the two projection maps commute with the action of ν.

We shall introduce further data and conditions in later chapters.

Chapter 3

Relative untwisted vertex operators

We shall present the notion of relative untwisted vertex operators, following our announcement [DL1]. We begin with some well-known structure (see e.g. [FLM3]), which we "relativize" to $\mathbf{h}_*$. Using the affinization of the abelian Lie algebra $\mathbf{h}$ together with the related Heisenberg algebra $\hat{\mathbf{h}}_{\mathbf{Z}}$ and the irreducible module $M(1)$, we construct the untwisted space V_L and the action of certain operators on it. The vacuum space Ω_* for the Heisenberg algebra $(\hat{\mathbf{h}}_*)_{\mathbf{Z}}$ associated with $\mathbf{h}_*$ plays a fundamental role. We define the relative vertex operators $Y_*(a, z)$ and more generally, $Y_*(v, z)$, using $\mathbf{h}_*$ systematically to modify the definition of the usual (unrelativized) vertex operators. Elementary properties of the operators are discussed. For the (important) degenerate case $\mathbf{h}_* = 0$, some of the notation simplifies, and the material in this chapter amounts to a review of the corresponding basic structure explained in [FLM3].

Viewing $\mathbf{h}$ as an abelian Lie algebra, we consider the corresponding affine Lie algebra

$$\hat{\mathbf{h}} = \mathbf{h} \otimes \mathbf{C}[t, t^{-1}] \oplus \mathbf{C}c \tag{3.1}$$

with structure defined by

$$[x \otimes t^m, y \otimes t^n] = \langle x, y\rangle m\delta_{m+n,0}c \quad \text{for } x, y \in \mathbf{h},\ m, n \in \mathbf{Z}, \tag{3.2}$$

$$[c, \hat{\mathbf{h}}] = 0. \tag{3.3}$$

(This notation and related notation below may be applied to any finite-dimensional abelian Lie algebra with a nonsingular symmetric form.)

By (2.12) we have

$$\hat{\mathbf{h}} = \mathbf{h}_* \otimes \mathbf{C}[t,t^{-1}] \oplus \mathbf{h}_*^\perp \otimes \mathbf{C}[t,t^{-1}] \oplus \mathbf{C}c. \tag{3.4}$$

Then $\hat{\mathbf{h}}$ has a $\mathbf{Z}$-gradation, *the weight gradation*, adapted to $\mathbf{h}_*$, given by:

$$\mathrm{wt}\,(x \otimes t^n) = 0, \quad \mathrm{wt}\,(y \otimes t^m) = -m, \quad \mathrm{wt}\, c = 0 \tag{3.5}$$

for $x \in \mathbf{h}_*$, $y \in \mathbf{h}_*^\perp$ and $m, n \in \mathbf{Z}$.

Remark 3.1: The case in which $\mathbf{h}_* = 0$ gives the ordinary weight gradation of $\hat{\mathbf{h}}$ as in [FLM3]. Also see (3.14) below.

We shall define a space V_L and natural actions of $\hat{\mathbf{h}}, \hat{L}$ and $\hat{\nu}$ on V_L. Set

$$\hat{\mathbf{h}}^+ = \mathbf{h} \otimes t\mathbf{C}[t], \quad \hat{\mathbf{h}}^- = \mathbf{h} \otimes t^{-1}\mathbf{C}[t^{-1}]. \tag{3.6}$$

The subalgebra

$$\hat{\mathbf{h}}_{\mathbf{Z}} = \hat{\mathbf{h}}^+ \oplus \hat{\mathbf{h}}^- \oplus \mathbf{C}c \tag{3.7}$$

of $\hat{\mathbf{h}}$ is a Heisenberg algebra, in the sense that its commutator subalgebra coincides with its center, which is one-dimensional. Consider the induced $\hat{\mathbf{h}}$-module, irreducible even under $\hat{\mathbf{h}}_{\mathbf{Z}}$,

$$M(1) = U(\hat{\mathbf{h}}) \otimes_{U(\mathbf{h}\otimes\mathbf{C}[t]\oplus\mathbf{C}c)} \mathbf{C} \simeq S(\hat{\mathbf{h}}^-) \quad \text{(linearly)}, \tag{3.8}$$

$\mathbf{h} \otimes \mathbf{C}[t]$ acting trivially on $\mathbf{C}$ and c acting as 1; $U(\cdot)$ denotes universal enveloping algebra and $S(\cdot)$ denotes symmetric algebra. (If c is made to act as multiplication by $k \in \mathbf{C}^\times$ instead, we obtain more generally a module $M(k)$, which will be needed only in Chapter 14.) The $\hat{\mathbf{h}}$-module $M(1)$ is $\mathbf{Z}$-graded so that $\mathrm{wt}\, 1 = 0$ (we write 1 for $1 \otimes 1$):

$$M(1) = \coprod_{n\in\mathbf{Z},\, n\geq 0} M(1)_n, \tag{3.9}$$

where $M(1)_n$ denotes the homogeneous subspace of weight n. The automorphism ν of L acts in a natural way on $\mathbf{h}$, on $\hat{\mathbf{h}}$ (fixing c) and on $M(1)$, preserving the gradations, and for $u \in \hat{\mathbf{h}}$ and $m \in M(1)$,

$$\nu(u \cdot m) = \nu(u) \cdot \nu(m). \tag{3.10}$$

Form the induced $\hat{L}$-module and $\mathbf{C}$-algebra

$$\begin{aligned} \mathbf{C}\{L\} &= \mathbf{C}[\hat{L}]/(\kappa_q - \omega_q)\mathbf{C}[\hat{L}] \\ &= \mathbf{C}[\hat{L}] \otimes_{\mathbf{C}[\langle\kappa_q\rangle]} \mathbf{C} \simeq \mathbf{C}[L] \quad \text{(linearly)}, \end{aligned} \tag{3.11}$$

where $\mathbf{C}[\cdot]$ denotes group algebra, ω_q is a primitive q^{th} root of unity in $\mathbf{C}^\times$ and κ_q acts as ω_q on $\mathbf{C}$. (Here $\langle\kappa_q\rangle$ is understood as an "abstract" group disjoint from $\mathbf{C}^\times$. Also, ω_q need not equal $e^{\frac{2\pi i}{q}}$ – the element of $\mathbf{C}^\times$ corresponding to κ_q.) For $a \in \hat{L}$, write $\iota(a)$ for the image of a in $\mathbf{C}\{L\}$. Then the action of $\hat{L}$ on $\mathbf{C}\{L\}$ and the product in $\mathbf{C}\{L\}$ are given by:

$$a \cdot \iota(b) = \iota(a)\iota(b) = \iota(ab), \tag{3.12}$$

$$\kappa_q \cdot \iota(b) = \omega_q \iota(b) \tag{3.13}$$

for $a, b \in \hat{L}$. We give $\mathbf{C}\{L\}$ the $\mathbf{C}$-gradation determined by:

$$\operatorname{wt} \iota(a) = \frac{1}{2}\langle \bar{a}', \bar{a}' \rangle \quad \text{for} \quad a \in \hat{L}. \tag{3.14}$$

The automorphism $\hat{\nu}$ of $\hat{L}$ (see (2.8)) acts canonically on $\mathbf{C}\{L\}$, preserving the gradation, in such a way that

$$\hat{\nu}\iota(a) = \iota(\hat{\nu}a) \quad \text{for} \quad a \in \hat{L}, \tag{3.15}$$

and we have

$$\hat{\nu}(\iota(a)\iota(b)) = \hat{\nu}(a \cdot \iota(b)) = \hat{\nu}(a) \cdot \hat{\nu}\iota(b) = \hat{\nu}\iota(a)\hat{\nu}\iota(b). \tag{3.16}$$

Also define a grading-preserving action of $\mathbf{h}$ on $\mathbf{C}\{L\}$ by:

$$h \cdot \iota(a) = \langle h', \bar{a} \rangle \iota(a) \tag{3.17}$$

for $h \in \mathbf{h}$ (so that $\mathbf{h}_*$ acts trivially). Then $\mathbf{h}$ acts as algebra derivations and

$$\hat{\nu}(h \cdot \iota(a)) = \nu(h) \cdot \hat{\nu}\iota(a). \tag{3.18}$$

We shall use a formal variable z (and later, commuting formal variables z, z_0, z_1, z_2, etc.). Define

$$z^h \cdot \iota(a) = z^{\langle h', \bar{a} \rangle}\iota(a) \tag{3.19}$$

for $h \in \mathbf{h}$. Then

$$\hat{\nu}(z^h \cdot \iota(a)) = z^{\nu(h)} \cdot \hat{\nu}\iota(a). \tag{3.20}$$

We shall mostly be interested in the actions of h and z^h on $\mathbf{C}\{L\}$ only for $h \in \mathbf{h}_*^\perp$.

Set

$$V_L = M(1) \otimes_{\mathbf{C}} \mathbf{C}\{L\} \simeq S(\hat{\mathbf{h}}^-) \otimes \mathbf{C}[L] \quad \text{(linearly)} \tag{3.21}$$

and give V_L the tensor product $\mathbf{C}$-gradation:

$$V_L = \coprod_{n \in \mathbf{C}} (V_L)_n. \tag{3.22}$$

We have $\mathrm{wt}\, \iota(1) = 0$, where we identify $\mathbf{C}\{L\}$ with $1 \otimes \mathbf{C}\{L\}$. Then $\hat{L}$, $\hat{\mathbf{h}}_{\mathbf{Z}}$, h, z^h $(h \in \mathbf{h})$ act naturally on V_L by acting on either $M(1)$ or $\mathbf{C}\{L\}$ as indicated above. In particular, c acts as 1 and $\mathbf{h}_*$ acts trivially. The automorphism $\hat{\nu}$ acts in a natural grading-preserving way on V_L, via $\nu \otimes \hat{\nu}$, and this action is compatible with the other actions:

$$\hat{\nu}(a \cdot v) = \hat{\nu}(a) \cdot \hat{\nu}(v) \tag{3.23}$$

$$\hat{\nu}(u \cdot v) = \nu(u) \cdot \hat{\nu}(v) \tag{3.24}$$

$$\hat{\nu}(z^h \cdot v) = z^{\nu(h)} \cdot \hat{\nu}(v) \tag{3.25}$$

for $a \in \hat{L}, u \in \hat{\mathbf{h}}, h \in \mathbf{h}, v \in V_L$.

It will be convenient later to have the notation

$$c(\alpha, \beta) = \omega_q^{c_0(\alpha,\beta)} \in \mathbf{C}^\times \tag{3.26}$$

for all $\alpha, \beta \in L$. Then

$$aba^{-1}b^{-1} = c(\bar{a}, \bar{b}) \tag{3.27}$$

for $a, b \in \hat{L}$, as operators on $\mathbf{C}\{L\}$ and on V_L, and

$$c(\nu\alpha, \nu\beta) = c(\alpha, \beta). \tag{3.28}$$

For $\alpha \in \mathbf{h}$, $n \in \mathbf{Z}$, we write $\alpha(n)$ for the operator on V_L determined by $\alpha \otimes t^n$. Set

$$\Omega_* = \{v \in V_L \mid h(n)v = 0 \ \text{ for } \ h \in \mathbf{h}_*, \ \ n > 0\}, \tag{3.29}$$

$$V_* = \mathrm{span}\{h(n)V_L \mid h \in \mathbf{h}_*, \ \ n < 0\}. \tag{3.30}$$

Then Ω_* is the vacuum space for the Heisenberg algebra $(\hat{\mathbf{h}}_*)_{\mathbf{Z}}$ (defined as in (3.7)) and we have

$$V_L = \Omega_* \oplus V_*. \tag{3.31}$$

In fact, we see from (3.21) that V_L has the decomposition

$$V_L = S(\hat{\mathbf{h}}_*^-) \otimes S((\hat{\mathbf{h}}_*^{\perp})^-) \otimes \mathbf{C}\{L\} \tag{3.32}$$

and so

$$\Omega_* = S((\hat{\mathbf{h}}_*^{\perp})^-) \otimes \mathbf{C}\{L\}, \tag{3.33}$$

$$V_* = \hat{\mathbf{h}}_*^- S(\hat{\mathbf{h}}_*^-) \otimes \Omega_*. \tag{3.34}$$

Here $\hat{\mathbf{h}}_*^-$ and $(\hat{\mathbf{h}}_*^{\perp})^-$ are defined as in (3.6). (If $\mathbf{h}_* = \mathbf{h}$, $\Omega_* = \mathbf{C}\{L\}$.) In terms of the general structure of modules for Heisenberg algebras, we know from [FLM3], Section 1.7, that the (well-defined) canonical linear map

$$\begin{array}{rcl} U((\hat{\mathbf{h}}_*)_{\mathbf{Z}}) \otimes_{(\hat{\mathbf{h}}_*^+ \oplus \mathbf{C}c)} \Omega_* & \to & V_L \\ u \otimes v & \mapsto & u \cdot v \end{array} \tag{3.35}$$

($u \in U((\hat{\mathbf{h}}_*)_{\mathbf{Z}}), v \in \Omega_*$) is an $(\hat{\mathbf{h}}_*)_{\mathbf{Z}}$-module isomorphism, and in particular, that the linear map

$$\begin{array}{rcl} M_*(1) \otimes_{\mathbf{C}} \Omega_* = U(\hat{\mathbf{h}}_*^-) \otimes_{\mathbf{C}} \Omega_* & \to & V_L \\ u \otimes v & \mapsto & u \cdot v \end{array} \tag{3.36}$$

($u \in U(\hat{\mathbf{h}}_*^-), v \in \Omega_*$) is an $(\hat{\mathbf{h}}_*)_{\mathbf{Z}}$-module isomorphism, Ω_* now being regarded as a trivial $(\hat{\mathbf{h}}_*)_{\mathbf{Z}}$-module, where $M_*(1)$ is the $(\hat{\mathbf{h}}_*)_{\mathbf{Z}}$-module defined by analogy with (3.8). The spaces Ω_* and V_* are $\mathbf{C}$-graded and are stable under the actions of $\hat{\mathbf{h}}_*^{\perp}$ (defined as in (3.1)), of $\hat{L}$ and of $\hat{\nu}$.

For $\alpha \in \mathbf{h}$, set

$$\alpha(z) = \sum_{n \in \mathbf{Z}} \alpha(n) z^{-n-1}. \tag{3.37}$$

We use a normal ordering procedure, indicated by open colons, which signify that the enclosed expression is to be reordered if necessary so that all the operators $\alpha(n)$ ($\alpha \in \mathbf{h}$, $n < 0$), $a \in \hat{L}$ are to be placed to the left of all the operators $\alpha(n)$, z^{α} ($\alpha \in \mathbf{h}$, $n \geq 0$) before the expression is evaluated. For $a \in \hat{L}$, set

$$Y_*(a, z) = {}^{\circ}_{\circ} e^{\int (\bar{a}'(z) - \bar{a}'(0) z^{-1})} a z^{\bar{a}'} {}^{\circ}_{\circ}, \tag{3.38}$$

using an obvious formal integration notation. (Note that the symbol $z^{\bar{a}'}$ could be replaced by $z^{\bar{a}}$ in this formula, in view of (3.19). If $\mathbf{h}_* = \mathbf{h}$, $Y_*(a, z) = a$.) Let

$$a \in \hat{L},\ \alpha_1, \ldots, \alpha_k \in \mathbf{h},\ n_1, \ldots, n_k \in \mathbf{Z}\ \ (n_i > 0)$$

and set

$$v = \alpha_1(-n_1)\cdots\alpha_k(-n_k)\otimes\iota(a) \\ = \alpha_1(-n_1)\cdots\alpha_k(-n_k)\cdot\iota(a)\in V_L. \tag{3.39}$$

We define

$$Y_*(v,z) = {\circ \atop \circ}\left(\frac{1}{(n_1-1)!}\left(\frac{d}{dz}\right)^{n_1-1}\alpha_1'(z)\right)\cdots \\ \cdot\left(\frac{1}{(n_k-1)!}\left(\frac{d}{dz}\right)^{n_k-1}\alpha_k'(z)\right)Y_*(a,z){\circ \atop \circ}. \tag{3.40}$$

This gives us a well-defined linear map

$$\begin{aligned} V_L &\to (\operatorname{End} V_L)\{z\} \\ v &\mapsto Y_*(v,z) = \sum_{n\in\mathbf{C}} v_n z^{-n-1}, \quad v_n\in\operatorname{End} V_L, \end{aligned} \tag{3.41}$$

where for any vector space W, we define $W\{z\}$ to be the vector space of W-valued formal series in z, with arbitrary complex powers of z allowed:

$$W\{z\} = \{\sum_{n\in\mathbf{C}} w_n z^n \mid w_n\in W\}. \tag{3.42}$$

We call $Y_*(v,z)$ the *untwisted vertex operator associated with* v, *defined relative to* $\mathbf{h}_*$. We shall be especially interested in these operators for $v\in\Omega_*$. Note that the component operators v_n of $Y_*(v,z)$ are defined in (3.41).

Remark 3.2: The case $\mathbf{h}_* = 0$ recovers the case of ordinary untwisted vertex operators as defined in [FLM3], Sections 8.4, 8.5. In this case, $\Omega_* = V_L$, $V_* = 0$, and Y_* is the operator Y, in the notation of [FLM3].

It is easy to check from the definition that the relative untwisted vertex operators $Y_*(v,z)$ for $v\in V_L$ have the following properties:

$$Y_*(1,z) = 1 \tag{3.43}$$

$$Y_*(v,z) = 0 \quad \text{if} \quad v\in V_* \tag{3.44}$$

$$Y_*(v,z)\iota(1)\in V_L[[z]] \text{ and } \lim_{z\to 0} Y_*(v,z)\iota(1) = v \text{ if } v\in\Omega_* \tag{3.45}$$

$$[(\hat{\mathbf{h}}_*)_{\mathbf{Z}}, Y_*(v,z)] = 0 \tag{3.46}$$

$$Y_*(a \cdot v, z) = aY_*(v, z) \tag{3.47}$$

for $a \in \hat{L}$ such that $\bar{a} \in L \cap \mathbf{h}_*$. (For a vector space W, $W[[z]]$ signifies the space of formal power series in z with coefficients in W.) Moreover, using (3.23)-(3.25) we see that

$$\hat{\nu} Y_*(v, z) \hat{\nu}^{-1} = Y_*(\hat{\nu} v, z). \tag{3.48}$$

By (3.46), the component operators v_n of $Y_*(v, z)$ preserve Ω_* and V_*, so that we get a well-defined linear map

$$\begin{aligned} V_L &\to (\operatorname{End} \Omega_*)\{z\} \\ v &\mapsto Y_*(v, z). \end{aligned} \tag{3.49}$$

We continue to denote the restriction of this map to Ω_* by Y_* :

$$\begin{aligned} \Omega_* &\to (\operatorname{End} \Omega_*)\{z\} \\ v &\mapsto Y_*(v, z). \end{aligned} \tag{3.50}$$

A basic property of the operators (3.38) and ultimately, of the operators (3.40), is: For $a, b \in \hat{L}$,

$$Y_*(a, z_1) Y_*(b, z_2) = {}^\circ_\circ Y_*(a, z_1) Y_*(b, z_2) {}^\circ_\circ (z_1 - z_2)^{\langle \bar{a}', \bar{b}' \rangle}, \tag{3.51}$$

where the binomial expression is to be expanded in nonegative integral powers of the second variable, z_2.

Remark 3.3: In our frequent use of formal series, it will typically be understood that a binomial expression such as that in (3.51) is to be expanded as in (3.51) – as a formal power series in the second variable.

Chapter 4

Quotient vertex operators

In this chapter we shall introduce what we call (in full) "quotient relative untwisted vertex operators" associated with a certain abelian subgroup A of $\hat{L}$. We use the structure developed in the last chapter to set up analogous structure on certain quotient spaces. A sublattice L_0 of L begins to play a role here. This sublattice will be important in later chapters, even in situations where $\mathbf{h}_* = 0$ and the relativization process degenerates; the reader interested only in this degenerate case may essentially skip this chapter, except for the definition of L_0.

Let L_0 be a sublattice of L such that

$$\text{rank}\, L_0 = \text{rank}\, L \tag{4.1}$$

(so that L_0 is nondegenerate) and L_0 is ν-invariant. Let A be a central subgroup of $\hat{L}_0$ $(= \{a \in \hat{L} \mid \bar{a} \in L_0\})$ satisfying

$$A \cap \langle \kappa_q \rangle = 1, \tag{4.2}$$

$$\hat{\nu} A = A \tag{4.3}$$

and

$$\bar{A} \subset \mathbf{h}_*, \tag{4.4}$$

where $\bar{A}$ is the (isomorphic) image of A under the homomorphism $^-$:

$$\bar{A} = \{\bar{a} \mid a \in A\} \simeq A. \tag{4.5}$$

Remark 4.1: Note that if $\mathbf{h}_* = 0$, then $A = 1$. In this situation, the considerations in this chapter will reduce to the ordinary case; recall Remarks 3.1, 3.2.

Form the induced right $\hat{L}$-module

$$\mathbf{C}\{L/\bar{A}\} = \mathbf{C} \otimes_{\mathbf{C}[A\langle\kappa_q\rangle]} \mathbf{C}[\hat{L}] \simeq \mathbf{C}[L/\bar{A}] \quad \text{(linearly)}, \tag{4.6}$$

where A acts trivially on $\mathbf{C}$ (on the right) and κ_q acts as multiplication by ω_q. The space $\mathbf{C}\{L/\bar{A}\}$ can be described alternatively as a quotient space of $\mathbf{C}\{L\}$,

$$\mathbf{C}\{L/\bar{A}\} = \mathbf{C}\{L\}/W_{\mathbf{C}\{L\}}, \tag{4.7}$$

where

$$W_{\mathbf{C}\{L\}} = \text{span}\,\{v - a \cdot v \mid v \in \mathbf{C}\{L\},\ a \in A\}. \tag{4.8}$$

For $b \in \hat{L}$, write $\iota(Ab)$ for the image of Ab in $\mathbf{C}\{L/\bar{A}\}$. Then $\iota(Ab) = \iota(b) + W_{\mathbf{C}\{L\}}$, $\hat{L}$ acts on $\mathbf{C}\{L/\bar{A}\}$ according to:

$$\iota(Ab) \cdot d = \iota(Abd) \tag{4.9}$$

for $b, d \in \hat{L}$ and we have

$$\iota(Ab) \cdot \kappa_q = \omega_q \iota(Ab). \tag{4.10}$$

Since A is central in $\hat{L}_0$, $\mathbf{C}\{L/\bar{A}\}$ is also a left $\hat{L}_0$-module under the action

$$b \cdot \iota(Ad) = \iota(Abd) \tag{4.11}$$

for $b \in \hat{L}_0$, $d \in \hat{L}$. Observe that A acts trivially and that $\mathbf{C}\{L/\bar{A}\}$ has a left $\hat{L}_0/A$-module structure under the action

$$bA \cdot \iota(Ad) = \iota(Abd) \tag{4.12}$$

for $b \in \hat{L}_0$, $d \in \hat{L}$.

Remark 4.2: If A is a central subgroup of $\hat{L}$, $\mathbf{C}\{L/\bar{A}\}$ is a left $\hat{L}$-module and

$$\mathbf{C}\{L/\bar{A}\} = \mathbf{C}[\hat{L}/A]/(\kappa_q - \omega_q)\mathbf{C}[\hat{L}/A]. \tag{4.13}$$

Furthermore, $\mathbf{C}\{L/\bar{A}\}$ is a $\mathbf{C}$-algebra with the product

$$\iota(Ab)\iota(Ad) = \iota(Abd). \tag{4.14}$$

In view of (4.4), we may give $\mathbf{C}\{L/\bar{A}\}$ the $\mathbf{C}$-gradation determined by:

$$\text{wt}\,\iota(Ab) = \text{wt}\,\iota(b) = \frac{1}{2}\langle \bar{b}', \bar{b}' \rangle \quad \text{for} \quad b \in \hat{L}. \tag{4.15}$$

The automorphism $\hat{\nu}$ acts on the space $\mathbf{C}\{L/\bar{A}\}$, preserving the gradation, so that

$$\hat{\nu}\iota(Ab) = \iota(\hat{\nu}Ab) \quad \text{for} \quad b \in \hat{L}, \tag{4.16}$$

and we have

$$\begin{aligned} \hat{\nu}(bA \cdot \iota(Ad)) &= \hat{\nu}(b \cdot \iota(Ad)) \\ &= \hat{\nu}(b) \cdot \hat{\nu}\iota(Ad) = \hat{\nu}(bA) \cdot \hat{\nu}\iota(Ad), \end{aligned} \tag{4.17}$$

$$\hat{\nu}(\iota(Ad) \cdot e) = \hat{\nu}\iota(Ad) \cdot \hat{\nu}(e) \tag{4.18}$$

for $b \in L_0$, $d, e \in \hat{L}$. We define a grading-preserving action of $\mathbf{h}$ on $\mathbf{C}\{L/\bar{A}\}$ by:

$$h \cdot \iota(Ab) = \langle h', \bar{b} \rangle \iota(Ab) \tag{4.19}$$

for $h \in \mathbf{h}$, $b \in \hat{L}$, so that

$$\hat{\nu}(h \cdot \iota(Ab)) = \nu(h) \cdot \hat{\nu}\iota(Ab). \tag{4.20}$$

Also set

$$z^h \cdot \iota(Ab) = z^{\langle h', \bar{b} \rangle} \iota(Ab) \tag{4.21}$$

for $h \in \mathbf{h}$. Then

$$\hat{\nu}(z^h \cdot \iota(Ab)) = z^{\nu(h)} \cdot \hat{\nu}\iota(Ab). \tag{4.22}$$

Now set

$$V_{L/\bar{A}} = M(1) \otimes_{\mathbf{C}} \mathbf{C}\{L/\bar{A}\} \simeq S(\hat{\mathbf{h}}^-) \otimes \mathbf{C}[L/\bar{A}] \quad \text{(linearly)}. \tag{4.23}$$

Then $\hat{L}$ acts on the right on $V_{L/\bar{A}}$ by acting on $\mathbf{C}\{L/\bar{A}\}$, and $\hat{L}_0$, $\hat{L}_0/A$, $\mathbf{h}_{\mathbf{Z}}$, h, z^h $(h \in \mathbf{h})$ act naturally on the left on $V_{L/\bar{A}}$ by acting on either $M(1)$ or $\mathbf{C}\{L/\bar{A}\}$ as indicated above. In particular, c acts as 1. As in (4.7), we can regard $V_{L/\bar{A}}$ as the quotient space of V_L by the subspace

$$W_{V_L} = \text{span}\,\{v - a \cdot v \mid v \in V_L,\ a \in A\} = M(1) \otimes W_{\mathbf{C}\{L\}}. \tag{4.24}$$

We give $V_{L/\bar{A}}$ the tensor product $\mathbf{C}$-gradation:

$$V_{L/\bar{A}} = \coprod_{n \in \mathbf{C}} (V_{L/\bar{A}})_n. \tag{4.25}$$

Then $\mathrm{wt}\,\iota(A) = \mathrm{wt}\,(\iota(1) + W_{V_L}) = 0$, where we identify $\mathbf{C}\{L/\bar{A}\}$ with $1 \otimes \mathbf{C}\{L/\bar{A}\}$. Also, $\hat{\nu}$ acts on $V_{L/\bar{A}}$ via $\nu \otimes \hat{\nu}$. This action preserves the grading, and formulas (3.23)-(3.25) hold here, this time with $a \in \hat{L}_0$, $v \in V_{L/\bar{A}}$; also,

$$\hat{\nu}(v \cdot b) = \hat{\nu}(v) \cdot \hat{\nu}(b) \quad \text{for} \quad b \in \hat{L}. \tag{4.26}$$

By analogy with (3.29) and (3.30) we set

$$(\Omega_*)_{L/\bar{A}} = \{v \in V_{L/\bar{A}} \mid h(n)v = 0 \ \text{ for } \ h \in \mathbf{h}_*, \ \ n > 0\}, \tag{4.27}$$

$$(V_*)_{L/\bar{A}} = \hat{\mathbf{h}}_*^- V_{L/\bar{A}}. \tag{4.28}$$

Then as in the earlier case, we have for instance

$$V_{L/\bar{A}} = (\Omega_*)_{L/\bar{A}} \oplus (V_*)_{L/\bar{A}}, \tag{4.29}$$

$$(\Omega_*)_{L/\bar{A}} = S((\hat{\mathbf{h}}_*^{\perp})^-) \otimes \mathbf{C}\{L/\bar{A}\}, \tag{4.30}$$

and these spaces are $\mathbf{C}$-graded and stable under $\hat{\mathbf{h}}_*^{\perp}$, $\hat{L}_0$ and $\hat{\nu}$, and under $\hat{L}$ acting on the right. (If $\mathbf{h}_* = \mathbf{h}$, $(\Omega_*)_{L/\bar{A}} = \mathbf{C}\{L/\bar{A}\}$.)

For a subset M of L (not necessarily a sublattice), we shall write

$$\hat{M} = \{b \in \hat{L} \mid \bar{b} \in M\} \tag{4.31}$$

and

$$\mathbf{C}\{M\} = \mathrm{span}\,\{\iota(b) \mid b \in \hat{M}\} \subset \mathbf{C}\{L\}, \tag{4.32}$$

$$V_M = M(1) \otimes \mathbf{C}\{M\} \subset V_L. \tag{4.33}$$

(Later we shall take M to be a coset of a lattice.) We may apply the considerations of Chapter 3 and the present chapter to $\hat{L}_0$ in place of $\hat{L}$. In particular, we have

$$\mathbf{C}\{L_0\} = \mathbf{C}[\hat{L}_0]/(\kappa_q - \omega_q)\mathbf{C}[\hat{L}_0], \tag{4.34}$$

$$V_{L_0} = M(1) \otimes \mathbf{C}\{L_0\}. \tag{4.35}$$

For $v \in V_{L_0}$, $Y_*(v,z)$ naturally acts on $V_{L/\bar{A}}$, on $(\Omega_*)_{L/\bar{A}}$ and on $(V_*)_{L/\bar{A}}$, and we have a linear map

$$\begin{aligned} V_{L_0} &\to (\mathrm{End}\, V_{L/\bar{A}})\{z\} \\ v &\mapsto Y_*(v,z) = \sum_{n \in \mathbf{C}} v_n z^{-n-1}, \quad v_n \in \mathrm{End}\, V_{L/\bar{A}}. \end{aligned} \tag{4.36}$$

By (3.47) and (4.4),

$$Y_*(a \cdot v, z) = Y_*(v, z) \tag{4.37}$$

for $a \in A$ and $v \in V_{L_0}$, as operators on $V_{L/\bar{A}}$, and therefore

$$Y_*(w, z) = 0 \quad \text{for} \quad w \in W_{V_{L_0}}. \tag{4.38}$$

Thus the linear map Y_* induces a linear map

$$\begin{aligned} V_{L_0/\bar{A}} &\to (\operatorname{End} V_{L/\bar{A}})\{z\} \\ v + W_{V_{L_0}} &\mapsto Y_*^A(v + W_{V_{L_0}}, z) = Y_*(v, z) \\ &\qquad = \sum_{n \in \mathbf{C}} (v + W_{V_{L_0}})_n z^{-n-1} \end{aligned} \tag{4.39}$$

($(v + W_{V_{L_0}})_n \in \operatorname{End} V_{L/\bar{A}}$) and we call $Y_*^A(v + W_{V_{L_0}}, z)$ the *quotient (relative untwisted) vertex operator* associated with v, with respect the group A. By (3.46) and (4.39) we see that the component operators $(v+W_{V_{L_0}})_n = v_n$ preserve $(\Omega_*)_{L/\bar{A}}$. If we restrict the map Y_*^A to $(\Omega_*)_{L_0/\bar{A}}$ ($(\Omega_*)_{L_0/\bar{A}}$ being defined as in (4.27)), then we have a well-defined linear map

$$\begin{aligned} (\Omega_*)_{L_0/\bar{A}} &\to (\operatorname{End}(\Omega_*)_{L/\bar{A}})\{z\} \\ v &\mapsto Y_*^A(v, z) = \sum_{n \in \mathbf{C}} v_n z^{-n-1}, \quad v_n \in \operatorname{End}(\Omega_*)_{L/\bar{A}}, \end{aligned} \tag{4.40}$$

and the relations (3.43), (3.45) and (3.48) hold for the quotient relative vertex operators $Y_*^A(v, z)$ for $v \in (\Omega_*)_{L_0/\bar{A}}$, namely:

$$Y_*^A(\iota(A), z) = 1 \tag{4.41}$$

$$\begin{aligned} &Y_*^A(v, z)\iota(A) \in (\Omega_*)_{L/\bar{A}}[[z]] \\ \text{and} \quad &\lim_{z \to 0} Y_*^A(v, z)\iota(A) = v \quad \text{if} \quad v \in (\Omega_*)_{L_0/\bar{A}} \end{aligned} \tag{4.42}$$

$$\hat{\nu} Y_*^A(v, z)\hat{\nu}^{-1} = Y_*^A(\hat{\nu} v, z). \tag{4.43}$$

Remark 4.3: If $\bar{A}$ spans $\mathbf{h}_*$ and L is positive definite, then each homogeneous subspace of $(\Omega_*)_{L/\bar{A}}$ is finite-dimensional.

Chapter 5

A Jacobi identity for relative untwisted vertex operators

We are now in a position to state and prove our "generalized Jacobi identity" for relative untwisted vertex operators. Even for $\mathbf{h}_* = 0$, this result generalizes the Jacobi identity in [FLM3] (Theorems 8.8.9 and 8.8.23) by removing all integrality restrictions on the "inner products" of lattice elements. Since the proof is very similar to that of Theorem 8.6.1 of [FLM3], we shall omit some computations, referring the reader to that proof for more details.

The main difference between the Jacobi identity presented in this chapter and the earlier one is that this identity incorporates certain algebraic functions, whose source is the nonintegrality of the "inner products" of lattice elements. We formulate a number of useful corollaries of the basic Jacobi identity, by passing to special cases and to the quotient structure of Chapter 4. Under a "rational compatibility" assumption, which asserts essentially that $\mathbf{h}_*$ and L are embedded compatibly in $\mathbf{h}$, we construct a finite abelian group G which gives rise to a G-grading of the space V_L. This grading serves to illuminate some of the data entering into the Jacobi identity. Under a suitable hypothesis, we establish an irreducibility result for the action of the vertex operators. Finally, we construct the natural Virasoro algebra associated with our structure and we give the basic properties, including the relations involving the differentiation of vertex operators, the grading of V_L and primary fields. These various structures and properties will be abstracted in later chapters, in various natural generalizations of the notions of vertex operator algebras and modules.

We recall some elementary material on formal calculus from [FLM3].

We shall use the basic generating function

$$(5.1) \qquad \delta(z) = \sum_{n \in \mathbf{Z}} z^n,$$

formally the expansion of the δ-function at $z = 1$. Its fundamental property is:

$$(5.2) \qquad f(z)\delta(z) = f(1)\delta(z) \quad \text{for} \quad f(z) \in \mathbf{C}[z, z^{-1}].$$

(This trivially holds for a monomial $f(z)$, and hence in general by linearity.) We shall use the following multivariable analogue of this principle: Given a family $X(m,n)$ $(m,n \in \mathbf{Z})$ of operators on a vector space V, set

$$(5.3) \qquad X(z_1, z_2) = \sum_{m,n \in \mathbf{Z}} X(m,n) z_1^m z_2^n$$

and assume that

$$(5.4) \qquad \lim_{z_1 \to z_2} X(z_1, z_2) = X(z_2, z_2) \quad \text{exists},$$

that is, when $X(z_1, z_2)$ is applied to any element of V and the variables are set equal, the coefficient of any monomial in z_2 is a finite sum in V. Then

$$(5.5) \quad X(z_1, z_2)\delta(z_1/z_2) = X(z_1, z_1)\delta(z_1/z_2) = X(z_2, z_2)\delta(z_1/z_2),$$

all these expressions existing (in the analogous sense). We shall also use the following formal version of Taylor's theorem: For a series of the form medskip

$$(5.6) \qquad Y(z_1, z_2) = \sum_{m,n \in \mathbf{C}} Y(m,n) z_1^m z_2^n \quad (Y(m,n) \in V),$$

we have

$$(5.7) \qquad e^{z_0 \frac{\partial}{\partial z_1}} Y(z_1, z_2) = Y(z_1 + z_0, z_2),$$

both sides of which are to be expanded in nonnegative integral powers of the formal variable z_0 in the obvious way – on the left we use the exponential series expansion, and on the right, each expression $(z_1 + z_0)^m$ is to be expanded as a formal power series in the second variable, z_0. Elementary consequences of these observations, such as Proposition

8.3.12 of [FLM3], enter into the proof below. The following general residue notation will be useful:

$$\operatorname{Res}_z\left(\sum v_n z^n\right) = v_{-1}. \tag{5.8}$$

Recall the notation $c(\alpha, \beta)$ from (3.26).

The result below expresses the first of several versions of our "generalized commutator relation" (5.10) and "generalized Jacobi identity" (5.11). The case $\mathbf{h}_* = 0$, $\langle \bar{a}, \bar{b} \rangle \in \mathbf{Z}$ coincides with Theorem 8.8.23 of [FLM3] (in which the factor with the exponent $\bar{a}$ belongs on the extreme right), and in particular, this result includes Theorem 8.6.1 of [FLM3].

Theorem 5.1 *Let $a, b \in \hat{L}$, $u^*, v^* \in S(\hat{\mathfrak{h}}^-) = M(1)$ and set*

$$\begin{aligned} u &= u^* \otimes \iota(a) \in V_L \\ v &= v^* \otimes \iota(b) \in V_L. \end{aligned} \tag{5.9}$$

Then as operators on V_L,

$$\begin{aligned} &(z_1 - z_2)^{n - \langle \bar{a}', \bar{b}' \rangle} Y_*(u, z_1) Y_*(v, z_2) \\ &- c(\bar{a}, \bar{b})(-1)^n (z_2 - z_1)^{n - \langle \bar{a}', \bar{b}' \rangle} Y_*(v, z_2) Y_*(u, z_1) \\ &= \operatorname{Res}_{z_0} z_0^{n - \langle \bar{a}', \bar{b}' \rangle} z_2^{-1} Y_*(Y_*(u, z_0)v, z_2) \left(\frac{z_1 - z_0}{z_2}\right)^{\bar{a}'} \delta\left(\frac{z_1 - z_0}{z_2}\right) \end{aligned} \tag{5.10}$$

for $n \in \mathbf{Z}$, or equivalently,

$$\begin{aligned} &z_0^{-1} \left(\frac{z_1 - z_2}{z_0}\right)^{-\langle \bar{a}', \bar{b}' \rangle} \delta\left(\frac{z_1 - z_2}{z_0}\right) Y_*(u, z_1) Y_*(v, z_2) \\ &- c(\bar{a}, \bar{b}) z_0^{-1} \left(\frac{z_2 - z_1}{z_0}\right)^{-\langle \bar{a}', \bar{b}' \rangle} \delta\left(\frac{z_2 - z_1}{-z_0}\right) Y_*(v, z_2) Y_*(u, z_1) \\ &= z_2^{-1} \delta\left(\frac{z_1 - z_0}{z_2}\right) Y_*(Y_*(u, z_0)v, z_2) \left(\frac{z_1 - z_0}{z_2}\right)^{\bar{a}'}, \end{aligned} \tag{5.11}$$

where all binomial expressions are to be expanded in nonnegative integral powers of the second variable.

Proof The equivalence being clear, we prove (5.10). Let $k, l \geq 1$ and let

$$a_1, \ldots, a_k, b_1, \ldots, b_l \in \hat{L}$$

subject to the conditions

$$a = a_1 \cdots a_k, \; b = b_1 \cdots b_l. \tag{5.12}$$

Set

$$\begin{aligned} A &= \exp\left(\sum_{i=1}^{k}\sum_{n\geq 1} \frac{\bar{a}_i'(-n)}{n} w_i^n\right) \cdot \iota(a) \in V_L[[w_1, ..., w_k]] \\ B &= \exp\left(\sum_{j=1}^{l}\sum_{n\geq 1} \frac{\bar{b}_j'(-n)}{n} x_j^n\right) \cdot \iota(b) \in V_L[[x_1, ..., x_l]]. \end{aligned} \tag{5.13}$$

Then the coefficients in the formal power series A and B span $S((\hat{\mathfrak{h}}_*^\perp)^-) \otimes \iota(a)$ and $S((\hat{\mathfrak{h}}_*^\perp)^-) \otimes \iota(b)$, respectively, as k, l, a_i, b_j vary subject to the constraint (5.12). Hence by (3.31), (3.33), (3.44) and (3.46) it suffices to prove the theorem with u and v replaced by A and B, respectively. We have

$$\begin{aligned} A &= \prod_{i=1}^{k} \exp\left(\sum_{n\geq 1} \frac{\bar{a}_i'(-n)}{n} w_i^n\right) \cdot \iota(a_1 \cdots a_k) \\ &= {}^\circ_\circ Y_*(a_1, w_1) \cdots Y_*(a_k, w_k) {}^\circ_\circ \iota(1) \\ &= Y_*(a_1, w_1) \cdots Y_*(a_k, w_k) \iota(1) \prod_{1\leq i<j\leq k} (w_i - w_j)^{-\langle \bar{a}_i', \bar{a}_j' \rangle}. \end{aligned} \tag{5.14}$$

Just as in the proof of Theorem 8.6.1 of [FLM3], we obtain

$$Y_*(A, z) = {}^\circ_\circ Y_*(a_1, z + w_1) \cdots Y_*(a_k, z + w_k) {}^\circ_\circ, \tag{5.15}$$

and similarly for $Y_*(B, z)$. We have

$${}^\circ_\circ Y_*(A, z_1) Y_*(B, z_2) {}^\circ_\circ = c(\bar{a}, \bar{b}) {}^\circ_\circ Y_*(B, z_2) Y_*(A, z_1) {}^\circ_\circ \tag{5.16}$$

and

$$\begin{aligned} Y_*(A, z_1) Y_*(B, z_2) &= {}^\circ_\circ Y_*(A, z_1) Y_*(B, z_2) {}^\circ_\circ \cdot \\ &\cdot \prod_{1\leq i\leq k, 1\leq j\leq l} (z_1 - z_2 + w_i - x_j)^{\langle \bar{a}_i', \bar{b}_j' \rangle}, \end{aligned} \tag{5.17}$$

where $(z_1 - z_2 + w_i - x_j)^{\langle \bar{a}_i', \bar{b}_j' \rangle}$ is understood as an expansion in nonnegative integral powers of z_2, w_i and x_j.

Fix a monomial

$$P = \prod_{1\leq i\leq k, 1\leq j\leq l} w_i^{r_i} x_j^{s_j} \quad (r_i, s_j \geq 0) \tag{5.18}$$

in the w_i and x_j. We may and do choose an integer $N \geq 0$ so large that the coefficient of P and of each monomial of lower total degree than P in

$$F_N = (z_1 - z_2)^N (z_1 - z_2)^{-\langle \bar{a}', \bar{b}' \rangle} \prod_{1 \leq i \leq k, 1 \leq j \leq l} (z_1 - z_2 + w_i - x_j)^{\langle \bar{a}'_i, \bar{b}'_j \rangle}$$
$$= (z_1 - z_2)^N \prod_{1 \leq i \leq k, 1 \leq j \leq l} \left(1 + \frac{w_i - x_j}{z_1 - z_2}\right)^{\langle \bar{a}'_i, \bar{b}'_j \rangle}$$

is a polynomial in $z_1 - z_2$. (We may use any $N \geq \deg P$.) Then this polynomial also equals the coefficient of the same monomial in

$$(z_1 - z_2)^N (z_2 - z_1)^{-\langle \bar{a}', \bar{b}' \rangle} \prod_{1 \leq i \leq k, 1 \leq j \leq l} (z_2 - z_1 + x_j - w_i)^{\langle \bar{a}'_i, \bar{b}'_j \rangle}$$
$$= (z_1 - z_2)^N \prod_{1 \leq i \leq k, 1 \leq j \leq l} \left(1 + \frac{x_j - w_i}{z_2 - z_1}\right)^{\langle \bar{a}'_i, \bar{b}'_j \rangle}.$$

Let $Y_P(z_1, z_2)$ denote the coefficient of P in

$$\begin{aligned} &Y_*(A, z_1) Y_*(B, z_2)(z_1 - z_2)^N (z_1 - z_2)^{-\langle \bar{a}', \bar{b}' \rangle} \\ &\quad = {}^{\circ}_{\circ} Y_*(A, z_1) Y_*(B, z_2) {}^{\circ}_{\circ} F_N. \end{aligned} \tag{5.19}$$

By (3.38) and (5.15) we see that

$$\begin{aligned} &Y_P(z_1, z_2) z_1^{-\bar{a}'} z_2^{-\bar{b}'} \in (\mathrm{End}\, V_L)[[z_1, z_1^{-1}, z_2, z_2^{-1}]], \\ &\lim_{z_1 \to z_2} Y_P(z_1, z_2) \quad \text{exists.} \end{aligned} \tag{5.20}$$

The coefficient of P in

$$Y_*(A, z_1) Y_*(B, z_2)(z_1 - z_2)^n (z_1 - z_2)^{-\langle \bar{a}', \bar{b}' \rangle}$$

is

$$Y_P(z_1, z_2)(z_1 - z_2)^{n-N}.$$

Similarly, reversing the roles of A and B and of z_1 and z_2 and noting that

$$ {}^{\circ}_{\circ} Y_*(B, z_2) Y_*(A, z_1) {}^{\circ}_{\circ} = c(\bar{a}, \bar{b})^{-1} {}^{\circ}_{\circ} Y_*(A, z_1) Y_*(B, z_2) {}^{\circ}_{\circ}, \tag{5.21}$$

we find that

$$Y_*(B, z_2) Y_*(A, z_1) = c(\bar{a}, \bar{b})^{-1} {}^{\circ}_{\circ} Y_*(A, z_1) Y_*(B, z_2) {}^{\circ}_{\circ}.$$

$$\cdot \prod_{1\leq i\leq k, 1\leq j\leq l} (z_2 - z_1 + x_j - w_i)^{\langle \bar{a}_i', \bar{b}_j'\rangle}.$$

Therefore the coefficient of P in

$$c(\bar{a}, \bar{b}) Y_*(B, z_2) Y_*(A, z_1)(-z_2 + z_1)^n (z_2 - z_1)^{-\langle \bar{a}', \bar{b}'\rangle}$$

is

$$Y_P(z_1, z_2)(-z_2 + z_1)^{n-N}.$$

Observe then that the coefficient of P in

$$\begin{aligned} &(z_1 - z_2)^{n-\langle \bar{a}', \bar{b}'\rangle} Y_*(A, z_1) Y_*(B, z_2) \\ &-c(\bar{a}, \bar{b})(-1)^n (z_2 - z_1)^{n-\langle \bar{a}', \bar{b}'\rangle} Y_*(B, z_2) Y_*(A, z_1) \end{aligned} \tag{5.22}$$

is

$$Y_P(z_1, z_2)\left((z_1 - z_2)^{n-N} - (-z_2 + z_1)^{n-N}\right).$$

Exactly as in the proof of Theorem 8.6.1 of [FLM3], this coefficient is

$$\begin{aligned} &\mathrm{Res}_{z_0} z_0^{n-N} z_2^{-1} Y_P(z_1, z_2) e^{-z_0 \frac{\partial}{\partial z_1}} \delta\left(\frac{z_1}{z_2}\right) \\ &= \mathrm{Res}_{z_0} z_0^{n-N} z_2^{-1} Y_P(z_2 + z_0, z_2) e^{-z_0 \frac{\partial}{\partial z_1}} \left(\left(\frac{z_1}{z_2}\right)^{\bar{a}'} \delta\left(\frac{z_1}{z_2}\right)\right). \end{aligned} \tag{5.23}$$

However, $Y_P(z_2 + z_0, z_2)$ is the coefficient of P in

$$\mathop{\circ}\limits_{\circ} Y_*(A, z_2 + z_0) Y_*(B, z_2) \mathop{\circ}\limits_{\circ} z_0^{N-\langle \bar{a}', \bar{b}'\rangle} \prod_{1\leq i\leq k, 1\leq j\leq l} (z_0 + w_i - x_j)^{\langle \bar{a}_i', \bar{b}_j'\rangle}.$$

We conclude that the coefficient of P in (5.22) is the coefficient of P in

$$\begin{aligned} &\mathrm{Res}_{z_0} z_0^{n-\langle \bar{a}', \bar{b}'\rangle} z_2^{-1} \mathop{\circ}\limits_{\circ} Y_*(A, z_2 + z_0) Y_*(B, z_2) \mathop{\circ}\limits_{\circ} \cdot \\ &\cdot \prod_{1\leq i\leq k, 1\leq j\leq l} (z_0 + w_i - x_j)^{\langle \bar{a}_i', \bar{b}_j'\rangle} e^{-z_0 \frac{\partial}{\partial z_1}} \left(\left(\frac{z_1}{z_2}\right)^{\bar{a}'} \delta\left(\frac{z_1}{z_2}\right)\right), \end{aligned} \tag{5.24}$$

which is independent of N. Since P is arbitrary, the expression (5.22) equals the expression (5.24).

On the other hand, from (5.14) and (5.15),

$$Y_*(A, z_0) B = \prod_{1\leq i\leq k, 1\leq j\leq l} (z_0 + w_i - x_j)^{\langle \bar{a}_i', \bar{b}_j'\rangle}.$$

$$\cdot \mathop{\circ}\limits_{\circ} Y_*(a_1, z_0 + w_1) \cdots Y_*(a_k, z_0 + w_k) Y_*(b_1, x_1) \cdots Y_*(b_l, x_l) \mathop{\circ}\limits_{\circ} \iota(1).$$

Thus by (5.15) with A replaced by $Y_*(A, z_0)B$,

$$Y_*(Y_*(A, z_0)B, z_2) = \prod_{1\le i\le k, 1\le j\le l} (z_0 + w_i - x_j)^{\langle \bar{a}'_i, \bar{b}'_j\rangle} \cdot$$
$$\cdot {}^{\circ}_{\circ} Y_*(a_1, z_2 + z_0 + w_1) \cdots Y_*(a_k, z_2 + z_0 + w_k) \cdot$$
$$\cdot Y_*(b_1, z_2 + x_1) \cdots Y_*(b_l, z_2 + x_l) {}^{\circ}_{\circ}$$
$$= {}^{\circ}_{\circ} Y_*(A, z_2 + z_0) Y_*(B, z_2) {}^{\circ}_{\circ} \prod_{1\le i\le k, 1\le j\le l} (z_0 + w_i - x_j)^{\langle \bar{a}'_i, \bar{b}'_j\rangle},$$

and this completes the proof. □

Note how this result simplifies if $\mathbf{h}_* = \mathbf{h}$. The following corollaries are immediate consequences of the definition (4.39) and the Jacobi identity (5.11).

Corollary 5.2 *Let A be a central subgroup of $\hat{L}_0$ satisfying (4.2) - (4.4). Let $a, b \in \hat{L}_0$, $u^*, v^* \in M(1)$ and set*

$$\begin{aligned} u &= u^* \otimes \iota(a) + W_{V_{L_0}} \in V_{L_0/\bar{A}} \\ v &= v^* \otimes \iota(b) + W_{V_{L_0}} \in V_{L_0/\bar{A}}. \end{aligned} \tag{5.25}$$

Then as operators on $V_{L/\bar{A}}$,

$$\begin{aligned} & z_0^{-1} \left(\frac{z_1 - z_2}{z_0}\right)^{-\langle \bar{a}', \bar{b}'\rangle} \delta\left(\frac{z_1 - z_2}{z_0}\right) Y_*^A(u, z_1) Y_*^A(v, z_2) \\ & - c(\bar{a}, \bar{b}) z_0^{-1} \left(\frac{z_2 - z_1}{z_0}\right)^{-\langle \bar{a}', \bar{b}'\rangle} \delta\left(\frac{z_2 - z_1}{-z_0}\right) Y_*^A(v, z_2) Y_*^A(u, z_1) \\ & = z_2^{-1} \delta\left(\frac{z_1 - z_0}{z_2}\right) Y_*^A(Y_*^A(u, z_0)v, z_2) \left(\frac{z_1 - z_0}{z_2}\right)^{\bar{a}'}. \end{aligned} \tag{5.26}$$

□

Corollary 5.3 *In the notation of Corollary 5.2, assume that $\langle \bar{a}, L\rangle \subset$* **Z**. *Then*

$$\begin{aligned} & z_0^{-1} \left(\frac{z_1 - z_2}{z_0}\right)^{\langle \bar{a}'', \bar{b}''\rangle} \delta\left(\frac{z_1 - z_2}{z_0}\right) Y_*^A(u, z_1) Y_*^A(v, z_2) \\ & - c(\bar{a}, \bar{b})(-1)^{\langle \bar{a}, \bar{b}\rangle} z_0^{-1} \left(\frac{z_2 - z_1}{z_0}\right)^{\langle \bar{a}'', \bar{b}''\rangle} \delta\left(\frac{z_2 - z_1}{-z_0}\right) Y_*^A(v, z_2) Y_*^A(u, z_1) \\ & = z_2^{-1} \delta\left(\frac{z_1 - z_0}{z_2}\right) Y_*^A(Y_*^A(u, z_0)v, z_2) \left(\frac{z_1 - z_0}{z_2}\right)^{-\bar{a}''}. \end{aligned} \tag{5.27}$$

□

(Recall the notation h'' from (2.13).)

We recover the Jacobi identity for ordinary untwisted vertex operators [FLM3] as follows:

Corollary 5.4 *Suppose that L is even, $L = L_0$, $\mathbf{h}_* = 0$ and $c(\alpha, \beta) = (-1)^{\langle \alpha,\beta \rangle}$ for all $\alpha, \beta \in L$. Then for $u, v \in V_L$, we have $Y_*(u, z) = Y(u, z)$ in the notation of [FLM3] and*

$$z_0^{-1}\delta\left(\frac{z_1 - z_2}{z_0}\right) Y(u, z_1)Y(v, z_2) - z_0^{-1}\delta\left(\frac{z_2 - z_1}{-z_0}\right) Y(v, z_2)Y(u, z_1)$$

$$= z_2^{-1}\delta\left(\frac{z_1 - z_0}{z_2}\right) Y(Y(u, z_0)v, z_2). \quad \square \tag{5.28}$$

We shall now introduce a finite abelian group G (which will be trivial in case $\mathbf{h}_* = 0$) equipped with a symmetric nondegenerate $\mathbf{Z}$-bilinear form, together with a G-gradation on $V_{L/\bar{A}}$, such that the parameters $\bar{a}'', \bar{b}''$ in the Jacobi identity (5.27) can be replaced by elements of G. For these purposes we make the following "rational compatibility" assumption, besides Assumptions 2.1-2.4 and (4.1)-(4.4): Setting

$$\mathbf{h}_{\mathbf{Q}} = \mathbf{Q} \otimes_{\mathbf{Z}} L, \tag{5.29}$$

we assume that

$$\mathbf{h}_{\mathbf{Q}} = (\mathbf{h}_* \cap \mathbf{h}_{\mathbf{Q}}) \oplus (\mathbf{h}_*^{\perp} \cap \mathbf{h}_{\mathbf{Q}}), \tag{5.30}$$

or equivalently,

$$\mathbf{h}_* \cap L \quad \text{spans} \quad \mathbf{h}_*, \tag{5.31}$$

or equivalently,

$$\mathbf{h}_*^{\perp} \cap L \quad \text{spans} \quad \mathbf{h}_*^{\perp}. \tag{5.32}$$

Then also

$$L' = \{\alpha' \mid \alpha \in L\} \subset \mathbf{Q}\text{-span}(\mathbf{h}_*^{\perp} \cap L) = \mathbf{h}_*^{\perp} \cap \mathbf{h}_{\mathbf{Q}}, \tag{5.33}$$

and L' and $\mathbf{h}_*^{\perp} \cap L$ are lattices of rank equal to $\dim \mathbf{h}_*^{\perp}$. Furthermore, there exists a smallest positive integer T such that

$$\langle \alpha', \beta' \rangle \in \frac{1}{T}\mathbf{Z}, \quad \langle \alpha, \beta \rangle \in \frac{1}{T}\mathbf{Z} \tag{5.34}$$

(so that $\langle \alpha'', \beta'' \rangle \in \frac{1}{T}\mathbf{Z}$) for $\alpha, \beta \in L$ and such that

$$\langle \gamma', \gamma' \rangle \in \frac{2}{T}\mathbf{Z} \quad \text{for} \quad \gamma \in L_0. \tag{5.35}$$

Then

$$Y_*^A(v,z) \in (\mathrm{End}\, V_{L/\bar{A}})[[z^{\frac{1}{T}}, z^{-\frac{1}{T}}]] \tag{5.36}$$

for $v \in V_{L_0/\bar{A}}$, so that we have a linear map

$$\begin{aligned} V_{L_0/\bar{A}} &\to (\mathrm{End}\, V_{L/\bar{A}})[[z^{\frac{1}{T}}, z^{-\frac{1}{T}}]] \\ v &\mapsto Y_*^A(v,z). \end{aligned} \tag{5.37}$$

Also,

$$\begin{aligned} V_{L_0/\bar{A}} &= \coprod_{n \in \frac{1}{T}\mathbf{Z}} (V_{L_0/\bar{A}})_n \\ V_{L/\bar{A}} &= \coprod_{n \in \frac{1}{2T}\mathbf{Z}} (V_{L/\bar{A}})_n. \end{aligned} \tag{5.38}$$

Now

$$L'' = \{\alpha'' \mid \alpha \in L\} \tag{5.39}$$

is a (rational) lattice spanning $\mathbf{h}_* \cap \mathbf{h}_{\mathbf{Q}}$. Using the general notation

$$S^\circ = \{\alpha \in \mathbf{h} \mid \langle \alpha, S\rangle \subset \mathbf{Z}\} \tag{5.40}$$

for a subset S of $\mathbf{h}$, we see that $(L'')^\circ \cap \mathbf{h}_*$ is also a lattice spanning $\mathbf{h}_* \cap \mathbf{h}_{\mathbf{Q}}$.

To ensure that the G-gradation (5.46) introduced below is well defined we also assume that

$$\langle L, \bar{A}\rangle \subset 2\mathbf{Z}. \tag{5.41}$$

Then

$$\bar{A} \subset L'' \cap 2(L'')^\circ. \tag{5.42}$$

Set

$$G = L''/H, \tag{5.43}$$

where

$$H = L'' \cap 2(L'')^\circ, \tag{5.44}$$

a lattice spanning $\mathbf{h}_* \cap \mathbf{h}_{\mathbf{Q}}$. Then G is a finite abelian group, and the lattice-isometry ν acts naturally as an automorphism of G. We define a $(\frac{1}{T}\mathbf{Z})/2\mathbf{Z}$-valued $\mathbf{Z}$-bilinear form $(\cdot,\cdot)$ on G by:

$$\begin{aligned} G \times G &\to \left(\tfrac{1}{T}\mathbf{Z}\right)/2\mathbf{Z} \\ (\alpha + H, \beta + H) &\mapsto (\alpha + H, \beta + H) = \langle \alpha, \beta\rangle + 2\mathbf{Z} \end{aligned} \tag{5.45}$$

for $\alpha, \beta \in L''$. Then this form is symmetric, nondegenerate and ν-invariant.

Now we define a G-gradation on $V_{L/\bar{A}}$. For $g \in G$ set

$$(5.46) \qquad (V_{L/\bar{A}})^g = \sum_{b \in \hat{L},\ \bar{b}''+H=g} (M(1) \otimes \iota(b) + W_{V_L}).$$

Then we have a decomposition

$$(5.47) \qquad V_{L/\bar{A}} = \coprod_{g \in G} (V_{L/\bar{A}})^g,$$

and similarly for $(\Omega_*)_{L/\bar{A}}$ and $(V_*)_{L/\bar{A}}$. This G-gradation is compatible with the earlier **C**-gradation (now a **Q**-gradation), and

$$(5.48) \qquad \begin{aligned} &Y_*^A(u,z)(V_{L/\bar{A}})^h \subset (V_{L/\bar{A}})^{g+h}[[z^{\frac{1}{T}}, z^{-\frac{1}{T}}]] \\ &Y_*^A(u,z)|_{(V_{L/\bar{A}})^h} = \sum_{n \equiv (g,h) \bmod \mathbf{Z}} u_n z^{-n-1} \quad \text{if } \langle L, L_0 \rangle \subset \mathbf{Z} \end{aligned}$$

(i.e., $n + 2\mathbf{Z} \equiv (g,h) \bmod \mathbf{Z}/2\mathbf{Z}$) for $g, h \in G$ and $u \in (V_{L_0/\bar{A}})^g$.

For $g \in G$ we define an operator $z^g \delta(z)$ on $V_{L/\bar{A}}$ by

$$(5.49) \qquad z^g \delta(z) \cdot u = z^{(g,h)} \delta(z) u$$

for $h \in G$ and $u \in (V_{L/\bar{A}})^h$; note that this is well defined. (Note that $z^g \delta(-z)$ is also well defined.) As usual, we shall sometimes take z to be a composite expression.

We have the following further corollary:

Corollary 5.5 *In the setting of Corollary 5.3, suppose that L_0 is an even sublattice of L such that*

$$(5.50) \qquad L \subset L_0^\circ, \quad \text{i.e.,} \quad \langle L, L_0 \rangle \subset \mathbf{Z},$$

and that q is even (see Chapter 2, (3.11), (3.26)) and

$$(5.51) \qquad c(\alpha, \beta) = (-1)^{\langle \alpha, \beta \rangle} \quad \textit{for} \quad \alpha, \beta \in L_0.$$

Let $g, h \in G$ and $u \in (V_{L_0/\bar{A}})^g, v \in (V_{L_0/\bar{A}})^h$. Then

$$(5.52) \qquad \begin{aligned} &z_0^{-1} \left(\frac{z_1 - z_2}{z_0}\right)^{(g,h)} \delta\left(\frac{z_1 - z_2}{z_0}\right) Y_*^A(u, z_1) Y_*^A(v, z_2) \\ &- z_0^{-1} \left(\frac{z_2 - z_1}{z_0}\right)^{(g,h)} \delta\left(\frac{z_2 - z_1}{-z_0}\right) Y_*^A(v, z_2) Y_*^A(u, z_1) \\ &= z_2^{-1} \delta\left(\frac{z_1 - z_0}{z_2}\right) Y_*^A(Y_*^A(u, z_0)v, z_2) \left(\frac{z_1 - z_0}{z_2}\right)^{-g}. \end{aligned}$$

□

It is natural to identify the smallest subspaces of $(\Omega_*)_{L/\bar{A}}$ invariant under the component operators of $Y_*^A(v,z)$ for all $v \in (\Omega_*)_{L_0/\bar{A}}$. In fact, the subspaces described below are irreducible modules for the generalized vertex operator algebra $(\Omega_*)_{L_0/\bar{A}}$, under appropriate additional hypotheses (see Chapter 6 below for a treatment of generalized vertex operator algebras and modules).

For $i \in L/L_0$, let λ_i be a coset representative in L, where we take $\lambda_0 = 0$, and consider the (finite) coset decomposition

$$L = \bigcup_{i \in L/L_0} (L_0 + \lambda_i). \tag{5.53}$$

Choose $\Lambda_i \in \hat{L}$ so that $\bar{\Lambda}_i = \lambda_i$. Then

$$\hat{L} = \bigcup_{i \in L/L_0} \hat{L}_0 \Lambda_i \tag{5.54}$$

is the corresponding coset decomposition of $\hat{L}$. For $i \in L/L_0$, set

$$\begin{aligned} V(i) = V_{L_0+\lambda_i} &= M(1) \otimes \mathbf{C}\{L_0 + \lambda_i\} \\ &\simeq S(\hat{\mathfrak{h}}^-) \otimes \mathbf{C}[L_0 + \lambda_i] \quad \text{(linearly)}, \end{aligned} \tag{5.55}$$

using the notation (4.32)-(4.33), and allowing the notation $\mathbf{C}[M]$ when M is not necessarily a group. Then we have

$$V_L = \coprod_{i \in L/L_0} V(i), \tag{5.56}$$

$$\Omega_* = \coprod_{i \in L/L_0} \Omega_*(i), \tag{5.57}$$

where $\Omega_*(i)$ is defined in the obvious way, and

$$V(0) = V_{L_0}. \tag{5.58}$$

We write

$$\Omega_*^A = (\Omega_*)_{L/\bar{A}}, \tag{5.59}$$

$$\Omega_*^A(i) = \Omega_*(i)/(\Omega_*(i) \cap W_{V_L}) \subset \Omega_*^A \tag{5.60}$$

for $i \in L/L_0$. Then

$$\Omega_*^A(i) = S((\hat{\mathfrak{h}}_*^\perp)^-) \otimes \mathbf{C}\{(L_0 + \lambda_i)/\bar{A}\} \tag{5.61}$$

(using an obvious extension of the notation (4.6)),

$$\Omega_*^A = \coprod_{i \in L/L_0} \Omega_*^A(i) \tag{5.62}$$

and

$$\Omega_*^A(0) = (\Omega_*)_{L_0/\bar{A}}. \tag{5.63}$$

Clearly, $\Omega_*^A(i)$ is invariant under the actions of the component operators of $Y_*^A(v,z)$ for $v \in \Omega_*^A(0)$. Standard techniques (cf. e.g. [FLM3], Proposition 4.4.2) give:

Proposition 5.6 *Assume that*

$$\bar{A} = L_0 \cap \mathbf{h}_*. \tag{5.64}$$

Then for each $i \in L/L_0$, $\Omega_*^A(i)$ is irreducible under the component operators of $Y_*^A(v,z)$ for $v \in \Omega_*^A(0)$. In particular (the case $\mathbf{h}_* = 0$), $V(i)$ is irreducible under the component operators of $Y(v,z)$ for $v \in V(0)$.

Proof Let W be a nonzero subspace of $\Omega_*^A(i)$ which is invariant under v_n for $n \in \mathbf{C}$ and $v \in \Omega_*^A(0)$. Since the span of these operators includes the affine algebra $\hat{\mathbf{h}}_*^\perp$, the general theory of modules for Heisenberg algebras (see Section 1.7 of [FLM3], cf. (3.35)-(3.36)) implies that

$$W = S((\hat{\mathbf{h}}_*^\perp)^-) \otimes \Omega$$

for some nonzero subspace Ω of $\mathbf{C}\{(L_0 + \lambda_i)/\bar{A}\}$. Since W is invariant under $\alpha(0)$ for $\alpha \in \mathbf{h}_*^\perp$, W contains a simultaneous eigenvector for these operators, and so there exist $b_1, ..., b_k \in \hat{L}_0$ and nonzero numbers $c_1, ..., c_k$ such that $\iota(Ab_1\Lambda_i), ..., \iota(Ab_k\Lambda_i)$ are linearly independent, $\bar{b}_1' = \cdots = \bar{b}_k'$ and

$$c_1\iota(Ab_1\Lambda_i) + \cdots + c_k\iota(Ab_k\Lambda_i) \in W.$$

Since $\bar{b}_j' - \bar{b}_l' = 0$ for $1 \le j, l \le k$, (5.64) shows that $\bar{b}_j - \bar{b}_l \in \bar{A}$. This implies that $b_j b_l^{-1} \in \kappa_q^r A$ for some $r \in \mathbf{Z}$. However, $\iota(A\kappa_q^r\Lambda_i) = \omega_q^r\iota(A\Lambda_i)$ and thus $\iota(Ab_j\Lambda_i) = \omega_q^r\iota(Ab_l\Lambda_i)$. This shows that $k = 1$, that is, $\iota(Ab_1\Lambda_i) \in W$. From (3.38) and (4.12) W is invariant under $\hat{L}_0/A$. Applying $b_1^{-1}A$ to $\iota(Ab_1\Lambda_i)$ we obtain $\iota(A\Lambda_i) \in W$, and we conclude that $W = \Omega_*^A(i)$. □

Now we construct a Virasoro algebra from a suitable relative untwisted vertex operator. The reader can refer to [FLM3] for a similar discussion and background.

Let $\{\beta_1, ..., \beta_d\}$ be an orthonormal basis of $\mathbf{h}_*^\perp$. Set

$$(5.65) \qquad \omega = \frac{1}{2}\sum_{i=1}^{d} \beta_i(-1)^2 = \frac{1}{2}\sum_{i=1}^{d} \beta_i(-1)^2 \cdot \iota(1) \in \Omega_* \subset V_L.$$

We define operators $L(n)$ on V_L as the following generating function coefficients:

$$(5.66) \qquad L(z) = Y_*(\omega, z) = \sum_{n\in\mathbf{Z}} L(n)z^{-n-2},$$

i.e.,

$$(5.67) \qquad L(n) = \omega_{n+1} \quad \text{for} \quad n \in \mathbf{Z}.$$

(recall (3.41)). (If $\mathbf{h}_* = \mathbf{h}$, $\omega = 0$ and $L(z) = 0$.)

First we describe the commutation action of $L(m)$ on $\alpha(n)$ for $\alpha \in \mathbf{h}$. By the Jacobi identity, as in [FLM3], formula (8.7.11)-(8.7.13), we have

$$(5.68) \quad [L(z_1), \alpha(z_2)] = \left(\frac{d}{dz_2}\alpha'(z_2)\right) z_2^{-1}\delta(z_1/z_2) - \alpha'(z_2) z_2^{-1}\frac{\partial}{\partial z_1}\delta(z_1/z_2)$$

$$(5.69) \qquad [L(m), \alpha(z_2)] = \left(z_2^{m+1}\frac{d}{dz_2} + (m+1)z_2^m\right)\alpha'(z_2)$$

$$(5.70) \qquad [L(m), \alpha(n)] = -n\alpha'(m+n)$$

for $\alpha \in \mathbf{h}, m, n \in \mathbf{Z}$.

We next give the following important connection between $L(-1)$ and differentiation:

Proposition 5.7 *For all $v \in V_L$,*

$$(5.71) \qquad Y_*(L(-1)v, z) = \frac{d}{dz}Y_*(v, z).$$

Proof : Use (5.70), the fact that

$$(5.72) \qquad L(-1)\cdot\iota(a) = \sum_{i=1}^{d} \beta_i(-1)\beta_i(0)\cdot\iota(a) = \bar{a}'(-1)\cdot\iota(a)$$

and the definition (3.40) of vertex operators. □

By the Jacobi identity and Proposition 5.7, for $v \in V_L$,

$$[L(z_1), Y_*(v, z_2)] = z_2^{-1}\left(\frac{d}{dz_2}Y_*(v, z_2)\right)\delta(z_1/z_2)$$

$$-z_2^{-1}Y_*(L(0)v, z_2)\frac{\partial}{\partial z_1}\delta(z_1/z_2)$$

$$(5.73) \qquad + z_2^{-1}\mathrm{Res}_{z_0}\sum_{n>0} Y_*(L(n)v, z_2)z_0^{-n-2}e^{-z_0\frac{\partial}{\partial z_1}}\delta(z_1/z_2).$$

Equating the coefficients of z_1^{-1} and changing z_2 to z, we get:

Proposition 5.8 *For all $v \in V_L$,*

$$(5.74) \qquad [L(-1), Y_*(v, z)] = \frac{d}{dz}Y_*(v, z) = Y_*(L(-1)v, z). \quad \square$$

We call a vector $v \in V_L$ a *weight vector* if v satisfies the condition:

$$(5.75) \qquad L(0)v = hv \quad \text{for some} \quad h \in \mathbf{C}$$

and we call h the *weight* of v. The weight vectors are precisely the homogeneous vectors with respect to our gradation, and

$$(5.76) \qquad v \in (V_L)_h \quad \text{for such } v.$$

Of course, $h \in \mathbf{Q}$. If v also satisfies the condition

$$(5.77) \qquad L(n)v = 0 \quad \text{for} \quad n > 0,$$

we call v a *lowest weight vector*. We have

$$(5.78) \qquad \mathrm{wt}\, v = 0 \quad \text{for} \quad v \in M_*(1),$$

and any $\iota(a)$ for $a \in \hat{L}$ is a lowest weight vector with weight $\frac{1}{2}\langle \bar{a}', \bar{a}' \rangle$; in particular,

$$(5.79) \qquad \mathrm{wt}\, \iota(a) = 0 \quad \text{if} \quad \bar{a}' = 0.$$

By (5.73) we have:

Proposition 5.9 *If $v \in V_L$ is a lowest weight vector with weight h, then $Y_*(v, z)$ is a "primary field of weight h," in the sense that*

$$(5.80) \qquad \begin{aligned}[L(z_1), Y_*(v, z_2)] &= z_2^{-1}\left(\frac{d}{dz_2}Y_*(v, z_2)\right)\delta(z_1/z_2)\\ &\quad -hz_2^{-1}Y_*(v, z_2)\frac{\partial}{\partial z_1}\delta(z_1/z_2),\end{aligned}$$

or equivalently,

$$(5.81) \qquad [L(n), Y_*(v, z)] = \left(z^{n+1}\frac{d}{dz} + h(n+1)z^n\right)Y_*(v, z)$$

for $n \in \mathbf{Z}$. $\square$

Taking $v = \omega$ in (5.73) and noting that

$$\begin{aligned} &L(n)\omega = 0, \quad n > 0 \ \text{ and } \ n \neq 2, \\ (5.82) \qquad &L(0)\omega = 2\omega, \\ &L(2)\omega = \frac{1}{2}(\dim \mathbf{h}_*^\perp)\iota(1), \end{aligned}$$

we have:

$$[L(z_1), L(z_2)] = z_2^{-1}\left(\frac{d}{dz_2}L(z_2)\right)\delta(z_1/z_2) - 2z_2^{-1}L(z_2)\frac{\partial}{\partial z_1}\delta(z_1/z_2)$$

$$(5.83) \qquad -\frac{1}{12}(\dim \mathbf{h}_*^\perp)z_2^{-1}\left(\frac{\partial}{\partial z_1}\right)^3 \delta(z_1/z_2).$$

Equating the coefficients of $z_1^{-m-2}z_2^{-n-2}$, we obtain

$$(5.84) \quad [L(m), L(n)] = (m-n)L(m+n) + \frac{1}{12}(m^3 - m)(\dim \mathbf{h}_*^\perp)\delta_{m+n,0}$$

for $m, n \in \mathbf{Z}$. We have:

Proposition 5.10 *If* $\mathrm{h}_* \neq \mathbf{h}$, *the operators* $L(n)$ $(n \in \mathbf{Z})$ *and 1 span a copy of the Virasoro algebra, and the operators* $L(n)$ *provide a representation of the Virasoro algebra on* V_L *with the canonical central element acting as multiplication by* $\dim \mathbf{h}_*^\perp$. □

Remark 5.11: Replacing Y_* and V_L by Y_*^A and $V_{L/\bar{A}}$, respectively, we get corresponding results for the quotient vertex operators.

Chapter 6

Generalized vertex operator algebras and their modules

Now that we have constructed some families of structures based on vertex operators satisfying a certain generalized Jacobi identity, we axiomatize crucial features of these structures in the notion of "generalized vertex operator algebra." We begin with our definition of this notion, and then we give some basic elementary properties and related definitions. We identify the precise sense in which the notion of (ordinary) vertex operator algebra, in the sense of [FLM3] and [FHL], is a special case. Then we formulate a theorem which clarifies the sense in which the structures developed in the earlier chapters give examples of generalized vertex operator algebras. After this, we carry out the analogous discussion for modules for generalized vertex operator algebras, giving our definition of the notion, followed by a discussion of related concepts and a theorem asserting that certain structures constructed earlier are examples of modules. We also explain reasonable generalizations of the notion of module. Certain of these generalizations – those based on a G-set S – will be very useful in later chapters. Finally, we discuss the special case of "vertex operator superalgebra."

It should be kept in mind that the notion of generalized vertex operator algebra will itself be generalized, in two stages, in this work, in order to incorporate wider families of examples: in Chapter 9, where we introduce "generalized vertex algebras," and then in Chapter 12, where we introduce the still more general "abelian intertwining algebras."

Let T be a positive integer. Let G be a finite abelian group and suppose that $(\cdot,\cdot)$ is a symmetric nondegenerate $(\frac{1}{T}\mathbf{Z})/2\mathbf{Z}$-valued $\mathbf{Z}$-

bilinear form on G :

$$(g,h) \in \left(\frac{1}{T}\mathbf{Z}\right)/2\mathbf{Z} \quad \text{for} \quad g,h \in G. \tag{6.1}$$

A generalized vertex operator algebra of *level* T associated with the group G and the form $(\cdot,\cdot)$ is a vector space with two gradations:

$$V = \coprod_{n\in\frac{1}{T}\mathbf{Z}} V_n = \coprod_{g\in G} V^g; \quad \text{for} \quad v \in V_n, \quad n = \operatorname{wt} v; \tag{6.2}$$

such that

$$V^g = \coprod_{n\in\frac{1}{T}\mathbf{Z}} V_n^g \quad (\text{where } V_n^g = V_n \bigcap V^g) \quad \text{for} \quad g \in G, \tag{6.3}$$

$$\dim V_n < \infty \qquad \text{for} \quad n \in \frac{1}{T}\mathbf{Z}, \tag{6.4}$$

$$V_n = 0 \quad \text{for} \quad n \quad \text{sufficiently small}, \tag{6.5}$$

equipped with a linear map

$$\begin{array}{rcl} V & \to & (\operatorname{End} V)[[z^{\frac{1}{T}}, z^{-\frac{1}{T}}]] \\ v & \mapsto & Y(v,z) = \displaystyle\sum_{n\in\frac{1}{T}\mathbf{Z}} v_n z^{-n-1} \quad (v_n \in \operatorname{End} V) \end{array} \tag{6.6}$$

and with two distinguished vectors $\mathbf{1} \in V_0^0$, $\omega \in V_2^0$, satisfying the following conditions for $g,h \in G$, $u,v \in V$ and $l \in \frac{1}{T}\mathbf{Z}$:

$$u_l V^h \subset V^{g+h} \quad \text{if} \quad u \in V^g; \tag{6.7}$$

$$u_l v = 0 \quad \text{if} \quad l \text{ is sufficiently large}; \tag{6.8}$$

$$Y(\mathbf{1},z) = 1; \tag{6.9}$$

$$Y(v,z)\mathbf{1} \in V[[z]] \quad \text{and} \quad \lim_{z\to 0} Y(v,z)\mathbf{1} = v; \tag{6.10}$$

$$Y(v,z)|_{V^h} = \sum_{n\equiv(g,h) \bmod \mathbf{Z}} v_n z^{-n-1} \quad \text{if} \quad v \in V^g \tag{6.11}$$

(i.e., $n + 2\mathbf{Z} \equiv (g,h) \bmod \mathbf{Z}/2\mathbf{Z}$);

$$z_0^{-1}\left(\frac{z_1 - z_2}{z_0}\right)^{(g,h)} \delta\left(\frac{z_1 - z_2}{z_0}\right) Y(u,z_1)Y(v,z_2)$$

$$-z_0^{-1}\left(\frac{z_2 - z_1}{z_0}\right)^{(g,h)} \delta\left(\frac{z_2 - z_1}{-z_0}\right) Y(v,z_2)Y(u,z_1)$$

$$(6.12) \qquad = z_2^{-1}\delta\left(\frac{z_1 - z_0}{z_2}\right) Y(Y(u, z_0)v, z_2)\left(\frac{z_1 - z_0}{z_2}\right)^{-g}$$

(the generalized Jacobi identity) if $u \in V^g$, $v \in V^h$, where

$$(6.13) \qquad \left(\frac{z_1 - z_0}{z_2}\right)^{-g}\delta\left(\frac{z_1 - z_0}{z_2}\right)\cdot w = \left(\frac{z_1 - z_0}{z_2}\right)^{-(g,k)}\delta\left(\frac{z_1 - z_0}{z_2}\right)w$$

for $w \in V^k$, $k \in G$ (note that all three terms in (6.12) are well defined, the symmetric form on G being defined modulo $2\mathbf{Z}$);

$$(6.14) \quad [L(m), L(n)] = (m-n)L(m+n) + \frac{1}{12}(m^3 - m)\delta_{m+n,0}(\mathrm{rank}\, V)$$

for $m, n \in \mathbf{Z}$, where

$$(6.15) \quad L(n) = \omega_{n+1} \text{ for } n \in \mathbf{Z}, \quad \text{i.e., } Y(\omega, z) = \sum_{n\in\mathbf{Z}} L(n)z^{-n-2}$$

and

$$(6.16) \qquad \mathrm{rank}\, V \in \mathbf{Q};$$

$$(6.17) \qquad L(0)v = nv = (\mathrm{wt}\, v)v \quad \text{for} \quad n \in \frac{1}{T}\mathbf{Z}, \quad v \in V_n;$$

$$(6.18) \qquad \frac{d}{dz}Y(v, z) = Y(L(-1)v, z).$$

This completes the definition. We denote the generalized vertex operator algebra just defined by

$$(6.19) \qquad (V, Y, \mathbf{1}, \omega, T, G, (\cdot, \cdot))$$

or briefly, by V. The expressions $Y(v, z)$ are called *(generalized) vertex operators*. Note the degenerate cases $\omega = 0$ and also $V = 0$ and $V = \mathbf{C}\mathbf{1}$.

Remark 6.1: Throughout this definition we may replace (g, h) for $g, h \in G$ by any one of its representatives in $\frac{1}{T}\mathbf{Z}$, such as (but not necessarily) the representative which is greater than or equal to 0 and less than 2. For instance, $z^g \cdot w = z^{(g,h)}w$ for $w \in V^h$ may be understood in this spirit. If we choose representatives in this way, then the form $(\cdot, \cdot)$ becomes bilinear only modulo $2\mathbf{Z}$. It is convenient to assume that the representatives (g, h) and (h, g) are equal for $g, h \in G$, and that $(g, 0) = 0$. These considerations can also be applied to modules, defined below.

The following are consequences of the definition: For $v \in V$,

$$[L(-1), Y(v,z)] = Y(L(-1)v, z) \tag{6.20}$$

$$[L(0), Y(v,z)] = Y(L(0)v, z) + zY(L(-1)v, z) \tag{6.21}$$

$$[L(1), Y(v,z)] = Y(L(1)v, z) + 2zY(L(0)v, z) + z^2Y(L(-1)v, z) \tag{6.22}$$

$$L(n)\mathbf{1} = 0 \quad \text{for} \quad n \geq -1 \tag{6.23}$$

$$L(-2)\mathbf{1} = \omega \tag{6.24}$$

$$L(0)\omega = 2\omega \tag{6.25}$$

$$Y(e^{z_0 L(-1)}v, z) = Y(v, z + z_0) \tag{6.26}$$

$$e^{z_0 L(-1)}Y(v,z)e^{-z_0 L(-1)} = Y(e^{z_0 L(-1)}v, z) \tag{6.27}$$

$$Y(v,z)\mathbf{1} = e^{zL(-1)}v. \tag{6.28}$$

Remark 6.2: For $g, h, k \in G$ and $v \in V^h$, $w \in V^k$,

$$z_1^g Y(v, z_2)w = Y(v, z_2)z_1^{g+(g,h)+m}w, \tag{6.29}$$

where $m \in 2\mathbf{Z}$ depends on g, h, k (recall Remark 6.1).

Remark 6.3: From (6.6), (6.17), (6.18) and (6.21) we find that if $v \in V$ is homogeneous with respect to the weight gradation, then v_n has weight $\operatorname{wt} v - n - 1$ as an operator.

Remark 6.4: The case $T = 1$, $G = 0$ in the definition of generalized vertex operator algebra gives precisely the definition of the vertex operator algebra $(V, Y, \mathbf{1}, \omega)$ ([FLM3], Section 8.10).

Remark 6.5: If V^0 satisfies the condition

$$V^0 = \coprod_{n \in \mathbf{Z}} V_n^0 \tag{6.30}$$

(recall (6.3)), then $(V^0, Y, \mathbf{1}, \omega)$ is a vertex operator algebra. In fact, this condition is always satisfied for the generalized vertex operator algebra constructed from an even lattice (see Theorem 6.7 below), for which we also have, more generally,

$$V^g = \coprod_{n \in -\frac{1}{2}(g,g)} V_n^g \quad \text{for} \quad g \in G \tag{6.31}$$

($-\frac{1}{2}(g,g)$ is a coset of $\mathbf{Z}$ in $\frac{1}{2T}\mathbf{Z}$). (Compare the $\mathbf{Q}$-grading, given by (4.15), with the bilinear from on G, given by (5.45), and use the fact that the lattice is even.) Note that the axiom (6.11) follows from the assertion (6.31) together with the other axioms for a generalized vertex operator algebra. This example illustrates how the G-grading can interact naturally with the $\mathbf{Q}$-grading via the quadratic form $\frac{1}{2}(g,g)$.

An *automorphism* of the generalized vertex operator algebra V of (6.19) consists of a pair (ρ,σ) where σ is an automorphism of G preserving the form $(\cdot,\cdot)$ and ρ is a linear automorphism of V preserving the $\frac{1}{T}\mathbf{Z}$-grading such that

$$\rho V^g = V^{\sigma g} \quad \text{for} \quad g \in G \tag{6.32}$$

$$\rho Y(v,z)\rho^{-1} = Y(\rho v, z) \quad \text{for} \quad v \in V \tag{6.33}$$

$$\rho \mathbf{1} = \mathbf{1} \tag{6.34}$$

$$\rho\omega = \omega. \tag{6.35}$$

The elements of G act naturally as automorphisms of V: For $g,h \in G$ and $v \in V^h$, we define

$$g \cdot v = e^{i\pi(g,h)}v. \tag{6.36}$$

Then g acts as the automorphism $(g, 1_G)$ of V. Moreover, G acts as a group of automorphisms of V.

Remark 6.6: The group G acts on each space V^h $(h \in G)$ as multiplication by the group character (i.e., the homomorphism $G \to \mathbf{C}^\times$)

$$g \mapsto e^{i\pi(g,h)}. \tag{6.37}$$

We have a natural homomorphism of G into the character group of G, sending h to the character (6.37), and the nondegeneracy of the form $(\cdot,\cdot)$ on G amounts to the assertion that this homomorphism is an injection, and hence an isomorphism. Thus the decomposition

$$V = \coprod_{h \in G} V^h \tag{6.38}$$

coincides with the character-space decomposition of V under the action of G.

Now we return to the setting of Chapter 5. Recall the gradings (5.38) and the spaces $\Omega_*^A = (\Omega_*)_{L/\bar{A}}$ (5.59) and $\Omega_*^A(0) = (\Omega_*)_{L_0/\bar{A}}$ (5.63). We have the decompositions

$$\Omega_*^A = \coprod_{n \in \frac{1}{2T}\mathbf{Z}} (\Omega_*^A)_n = \coprod_{g \in G} (\Omega_*^A)^g \tag{6.39}$$

$$\Omega_*^A(0) = \coprod_{n \in \frac{1}{T}\mathbf{Z}} (\Omega_*^A(0))_n = \coprod_{g \in G} (\Omega_*^A(0))^g, \tag{6.40}$$

where T is the smallest positive integer satisfying (5.34)-(5.35) and where G is the finite abelian group given by (5.43), defining the grading given by (5.46)-(5.47) and equipped with the $\mathbf{Z}$-bilinear form given by (5.45).

Much of the structure developed so far is summarized as follows:

Theorem 6.7 *Let L be a positive definite lattice satisfying Assumptions 2.1-2.4. Let L_0 be a ν-invariant even sublattice of L satisfying (4.1), (5.50) and (5.51) (with q even). Let A be a central subgroup of $\hat{L}_0$ such that (4.2)-(4.4), (5.41) hold and such that $\bar{A}$ spans $\mathbf{h}_*$. We also assume the rational compatibility condition (5.30). Then*

$$(\Omega_*^A(0), Y_*^A, \iota(A), \omega \otimes \iota(A), T, G, (\cdot,\cdot)) \tag{6.41}$$

is a generalized vertex operator algebra with rank equal to $\dim \mathbf{h}_*^\perp$, *and $\hat{\nu}$ acts naturally as an automorphism. In particular, if L is even and $L = L_0$, then*

$$((\Omega_*)_{L/\bar{A}}, Y_*^A, \iota(A), \omega \otimes \iota(A), T, G, (\cdot,\cdot)) \tag{6.42}$$

is a generalized vertex operator algebra. □

Remark 6.8: We know that G and the group $\langle\hat{\nu}\rangle$ generated by $\hat{\nu}$ both act as automorphisms of V. In fact, their semidirect product $\langle\hat{\nu}\rangle \ltimes G$ acts on V as well, since

$$\hat{\nu} g \hat{\nu}^{-1} = \nu g \quad \text{for} \quad g \in G \tag{6.43}$$

as operators on both G and V. More generally, for any group K of automorphisms of the generalized vertex operator algebra (6.19), the group $K \ltimes G$ acts as a group of automorphisms.

Remark 6.9: If L is even, $L = L_0$ and $\mathbf{h}_* = 0$ in Theorem 6.7, then $A = 1$, $T = 1$, $G = 0$, $(\Omega_*)_{L/\bar{A}} = V_L$ and $Y_*^A = Y$ in the notation of [FLM3], and Theorem 6.7 says that $(V_L, Y, \mathbf{1}, \omega)$ is a vertex operator algebra ([FLM3], Section 8.10). Also note the case $\mathbf{h}_* = \mathbf{h}$.

The following is the appropriate notion of module:

Let $V = (V, Y, \mathbf{1}, \omega, T, G, (\cdot,\cdot))$ be a generalized vertex operator algebra. A *module* W for V is a $\mathbf{Q}$-graded and G-graded vector space:

$$W = \coprod_{n\in\mathbf{Q}} W_n = \coprod_{g\in G} W^g; \quad \text{for} \quad w \in W_n, \quad n = \mathrm{wt}\, w; \tag{6.44}$$

such that

$$W^g = \coprod_{n\in\mathbf{Q}} W_n^g \quad (\text{where } W_n^g = W_n \bigcap W^g) \quad \text{for} \quad g \in G, \tag{6.45}$$

$$\dim W_n < \infty \qquad \text{for} \quad n \in \mathbf{Q}, \tag{6.46}$$

$$W_n = 0 \qquad \text{for} \quad n \quad \text{sufficiently small}, \tag{6.47}$$

equipped with a linear map

$$\begin{array}{rcl} V & \to & (\mathrm{End}\, W)[[z^{\frac{1}{T}}, z^{-\frac{1}{T}}]] \\ v & \mapsto & Y(v,z) = \displaystyle\sum_{n\in\frac{1}{T}\mathbf{Z}} v_n z^{-n-1} \quad (v_n \in \mathrm{End}\, W) \end{array} \tag{6.48}$$

(note that the sum is over $\frac{1}{T}\mathbf{Z}$, not $\mathbf{Q}$), such that "all the defining properties of a generalized vertex operator algebra that make sense hold." That is, for $g, h \in G$, $u, v \in V$, $w \in W$ and $l \in \frac{1}{T}\mathbf{Z}$:

$$u_l W^h \subset W^{g+h} \quad \text{if} \quad u \in V^g; \tag{6.49}$$

$$u_l w = 0 \quad \text{if} \quad l \ \text{is sufficiently large}; \tag{6.50}$$

$$Y(\mathbf{1}, z) = 1; \tag{6.51}$$

$$Y(v,z)|_{W^h} = \sum_{n\equiv(g,h) \bmod \mathbf{Z}} v_n z^{-n-1} \quad \text{if} \quad v \in V^g \tag{6.52}$$

(i.e., $n + 2\mathbf{Z} \equiv (g,h) \bmod \mathbf{Z}/2\mathbf{Z}$);

$$\begin{aligned} & z_0^{-1}\left(\frac{z_1-z_2}{z_0}\right)^{(g,h)} \delta\left(\frac{z_1-z_2}{z_0}\right) Y(u,z_1)Y(v,z_2) \\ & - z_0^{-1}\left(\frac{z_2-z_1}{z_0}\right)^{(g,h)} \delta\left(\frac{z_2-z_1}{-z_0}\right) Y(v,z_2)Y(u,z_1) \\ & = z_2^{-1}\delta\left(\frac{z_1-z_0}{z_2}\right) Y(Y(u,z_0)v,z_2)\left(\frac{z_1-z_0}{z_2}\right)^{-g} \end{aligned} \tag{6.53}$$

(the generalized Jacobi identity for operators on W) if $u \in V^g, v \in V^h$, where $Y(u, z_0)$ is an operator on V;

$$[L(m), L(n)] = (m-n)L(m+n) + \frac{1}{12}(m^3-m)\delta_{m+n,0}(\mathrm{rank}\, V) \tag{6.54}$$

for $m, n \in \mathbf{Z}$, where

$$L(n) = \omega_{n+1} \text{ for } n \in \mathbf{Z}, \text{ i.e., } Y(\omega, z) = \sum_{n \in \mathbf{Z}} L(n) z^{-n-2}; \tag{6.55}$$

$$L(0)w = nw = (\mathrm{wt}\, w)w \text{ for } w \in W_n \ (n \in \mathbf{Q}); \tag{6.56}$$

$$\frac{d}{dz} Y(v, z) = Y(L(-1)v, z), \tag{6.57}$$

where $L(-1)$ is the operator on V. This completes the definition. We denote this module by

$$(W, Y) \tag{6.58}$$

or by W. The formal series $Y(v, z)$ are called *(generalized) vertex operators*. Recall Remark 6.1. One has the usual module notions of homomorphism, direct sum, irreducibility, and so on.

Remark 6.10: A generalized vertex operator algebra is clearly a module for itself. We call it the *adjoint module.*

As in the case of algebras, we have the following consequences of the definition:

$$[L(-1), Y(v, z)] = Y(L(-1)v, z) \tag{6.59}$$

$$[L(0), Y(v, z)] = Y(L(0)v, z) + zY(L(-1)v, z). \tag{6.60}$$

$$[L(1), Y(v, z)] = Y(L(1)v, z) + 2zY(L(0)v, z) + z^2 Y(L(-1)v, z) \tag{6.61}$$

$$Y(e^{z_0 L(-1)} v, z) = Y(v, z + z_0) \tag{6.62}$$

$$e^{z_0 L(-1)} Y(v, z) e^{-z_0 L(-1)} = Y(e^{z_0 L(-1)} v, z). \tag{6.63}$$

Remark 6.11: Formulas (6.48), (6.56), (6.57) and (6.60) imply that if $v \in V$ is homogeneous with respect to the weight gradation, then v_n has weight $\mathrm{wt}\, v - n - 1$ as an operator on W.

Remark 6.12: If $T = 1$ and $G = 0$ as in Remark 6.4, the notion of V-module reduces to the notion of module for the vertex operator algebra $(V, Y, \mathbf{1}, \omega)$, in the sense of the definition in [FLM3].

Remark 6.13: Recalling Remark 6.5, if (6.30) holds, we find that (W^g, Y) is a module for the vertex operator algebra $(V^0, Y, \mathbf{1}, \omega)$ for $g \in G$.

Theorem 6.14 *Suppose that all the assumptions in Theorem 6.7 hold. Then* (Ω_*^A, Y_*^A) *and* $(\Omega_*^A(i), Y_*^A)$ *for* $i \in L/L_0$ *(recall (5.61), (5.62)) are modules for the generalized vertex operator algebra* $\Omega_*^A(0)$. *If (5.64) is also satisfied, then each* $\Omega_*^A(i)$ *is irreducible.* □

Remark 6.15: We have the following natural generalization of the notion of module for a generalized vertex operator algebra V: Let S be a G-set (i.e., a set on which G acts), with the action of G on S denoted by $+$, equipped with a $(\frac{1}{T}\mathbf{Z})/2\mathbf{Z}$-valued function $(\cdot,\cdot)$ on $G \times S$ such that

$$(g+h, k+s) = (g,k) + (g,s) + (h,k) + (h,s) \tag{6.64}$$

for $g, h, k \in G$ and $s \in S$. A *module* W (in the new sense) for V associated with S and $(\cdot,\cdot)$ is a $\mathbf{Q}$-graded and S-graded vector space such that all the assumptions in the definition of module hold for W with obvious modifications. We denote this module by

$$W = (W, Y, S, (\cdot,\cdot)). \tag{6.65}$$

(Actually, for many purposes, the values of the function $(\cdot,\cdot)$ can be reduced mod $\mathbf{Z}$ or can be directly assumed to lie in $(\frac{1}{T}\mathbf{Z})/\mathbf{Z}$ rather than $(\frac{1}{T}\mathbf{Z})/2\mathbf{Z}$, but when we study intertwining operators later, for example, we shall need the $(\frac{1}{T}\mathbf{Z})/2\mathbf{Z}$-values.) The comments above concerning modules carry over appropriately. It is worth noting that even in the case of an ordinary vertex operator algebra V (the case $T = 1$, $G = 0$; recall Remark 6.4), it is sometimes appropriate to equip V-modules with this additional structure, which in this case amounts precisely to a distinguished decomposition of a V-module W into submodules W^s for $s \in S$. If W is irreducible, then S can of course be taken to be a one-element set. This structure, even in case S has only one element, will turn out to be useful when we study intertwining operators.

Remark 6.16: In the definition of module, one can also generalize the assumption of a $\mathbf{Q}$-grading on W to the assumption of a $\mathbf{C}$-grading; this also entails replacing (6.47) by the condition that $W_n = 0$ for $n \in \mathbf{C}$ sufficiently small in the sense of modification by a rational number, that is, the condition that for $m \in \mathbf{C}$, $W_{m-n} = 0$ for $n \in \mathbf{Q}$ sufficiently large.

Remark 6.17: The case $T = 2$ and $G = \mathbf{Z}/2\mathbf{Z}$, with $(1,1) = 1 + 2\mathbf{Z}$ (recall (6.1)) in the definition of generalized vertex operator algebra gives a notion of *vertex operator superalgebra* $(V, Y, \mathbf{1}, \omega, \mathbf{Z}/2\mathbf{Z})$. Note that the form $(\cdot,\cdot)$ is $(\mathbf{Z}/2\mathbf{Z})$-valued, and that it amounts simply to multiplication in $\mathbf{Z}/2\mathbf{Z}$. Note also that for $v \in V$, $Y(v,z)$ involves only integral powers of z, by (6.11). Let us write $|v| = 0$ if $v \in V^0$ and $|v| = 1$ if $v \in V^1$. Then for G-homogeneous vectors $u, v \in V$, the generalized Jacobi identity becomes the *super-Jacobi identity:*

$$(6.66)\qquad \begin{aligned} &z_0^{-1}\delta\left(\frac{z_1 - z_2}{z_0}\right) Y(u,z_1)Y(v,z_2) \\ &-(-1)^{|u||v|} z_0^{-1}\delta\left(\frac{z_2 - z_1}{-z_0}\right) Y(v,z_2)Y(u,z_1) \\ &= z_2^{-1}\delta\left(\frac{z_1 - z_0}{z_2}\right) Y(Y(u,z_0)v,z_2). \end{aligned}$$

In Theorem 6.7, if $T = 2$, $\langle L'', L''\rangle \subset \mathbf{Z}$, $G \simeq \mathbf{Z}/2\mathbf{Z}$, then $(1,1) = 1+2\mathbf{Z}$ (recall from (5.45) that the form $(\cdot,\cdot)$ on G is nondegenerate), and so $(\Omega_*^A(0), Y_*^A, \iota(A), \omega\otimes\iota(A), \mathbf{Z}/2\mathbf{Z})$ is an example of a vertex operator superalgebra. See Remarks 9.21 and 12.38 below for further discussion.

Chapter 7

Duality for generalized vertex operator algebras

This chapter is patterned after Section 3 of [FHL]. We shall present three important properties of a generalized vertex operator algebra – generalized rationality, generalized commutativity and generalized associativity – and we shall show that they may be used in place of the generalized Jacobi identity in the definition of generalized vertex operator algebra. These properties are aspects of "duality," in the terminology of conformal field theory. At the end of this chapter, we comment that most of the discussion also applies to modules. In the next chapter, we shall also show how generalized commutativity leads to one-dimensional representations of braid groups.

The equivalence between the Jacobi identity and the duality properties is fundamental to the theory of vertex operator algebras, providing a link between the Jacobi identity and more geometric concepts. Generalized rationality asserts that suitable "matrix coefficients" of products and other natural combinations (iterates) of vertex operators are formal series which, after multiplication by certain formal series expansions of certain algebraic functions, become the formal series expansions of rational functions of the formal variables, expansions which are valid in certain domains. Generalized commutativity then asserts that these rational functions are equal for the products of two vertex operators in opposite orders, and generalized associativity asserts that this rational function agrees with that arising from a suitable iterate of the two vertex operators. Different expansion domains must be considered at once; it is only the rational functions and not the formal series which are equal, in the commutativity and associativity relations. In the special case of (ordinary) vertex operator algebras, the auxiliary algebraic functions needed to produce the rational functions become trivial.

Many of the arguments in this chapter involve useful principles, and even for the case of (ordinary) vertex operator algebras, the treatment presented in this chapter contains a little more information than does the corresponding treatment of the analogous material in [FLM3] and [FHL]. In fact, the material in this chapter may be usefully specialized to the case in which the generalized vertex operator algebra is a vertex operator algebra; some of the notation simplifies, and the resulting treatment is an essentially self-contained replacement, with a bit of improvement, of the corresponding treatments in [FLM3] and [FHL]. After deriving from the Jacobi identity a "generalized commutator relation" and a "generalized associator relation," which reduce to corresponding "commutator" and "associator" relations in the case of vertex operator algebras, we formulate and establish the generalized rationality, commutativity and associativity relations. In particular, the generalized associativity relation is a manifestation of the "associativity of the operator product expansion" in conformal field theory. These formal-series results immediately imply corresponding corollaries on convergent series expansions of the relevant rational functions. Then we show conversely that the generalized rationality, commutativity and associativity relations imply the Jacobi identity and hence can be taken to replace the Jacobi identity in the axioms for a generalized vertex operator algebra. Generalized rationality and commutativity are extended to products of more than two vertex operators, by an argument which slightly simplifies the corresponding argument in [FHL] (when restricted to the special case of vertex operator algebras); this result will be used in the next chapter to construct canonical one-dimensional braid group representations. The final result in this chapter establishes that generalized associativity, and hence the Jacobi identity, follows from generalized commutativity alone, in the presence of certain relatively mild assumptions. It is remarked at the end of the chapter that a large part of the discussion of duality in this chapter also applies, with obvious modifications, to modules for generalized vertex operator algebras.

Our formulations of the generalized rationality properties incorporate a small but subtle and important addition to the statements of the corresponding rationality properties (for the special case of vertex operator algebras) in [FLM3] and [FHL]: Here we include, in the statements of the generalized rationality of both products (of two or more vertex operators) and iterates the assertion that the orders of certain poles in the occurring rational functions are bounded by numbers depending only on certain pairs of vectors and in particular are independent of the other vectors entering into the statements. This refinement of the notion of generalized rationality enables us for example to conclude directly that in the presence of certain mild conditions, generalized rationality and commutativity for two vertex operators implies generalized rationality and commutativity for an arbitrary number of vertex operators. These considerations are just as relevant for ordinary vertex operator algebras ([FLM3] and [FHL]) as for generalized vertex operator algebras.

The reader should keep in mind that the results and proofs in this chapter will be adapted below to the more general settings of generalized vertex algebras and abelian intertwining algebras. The proofs will usually not be repeated, since they would be essentially the same as those in this chapter. Also, the reader should note how the development in this chapter simplifies when carried out for the special case of vertex operator superalgebras (recall Remark 6.17); here the form $(\cdot,\cdot)$ on G takes integral values modulo $2\mathbf{Z}$, eliminating any monodromy (multiple-valuedness) in the Jacobi identity.

In addition to the basic properties of the δ-function presented at the beginning of Chapter 5, we shall also need:

$$(7.1)\quad z_1^{-1}\left(\frac{z_2+z_0}{z_1}\right)^r\delta\left(\frac{z_2+z_0}{z_1}\right)=z_2^{-1}\left(\frac{z_1-z_0}{z_2}\right)^{-r}\delta\left(\frac{z_1-z_0}{z_2}\right)$$

$$(7.2)\quad z_0^{-1}\delta\left(\frac{z_1-z_2}{z_0}\right)-z_0^{-1}\delta\left(\frac{z_2-z_1}{-z_0}\right)=z_2^{-1}\delta\left(\frac{z_1-z_0}{z_2}\right)$$

for $r\in\mathbf{C}$ (see [FLM3], Propositions 8.8.22 and 8.8.15).

Remark 7.1: We shall often use the notation (g,h) for $g,h\in G$ in this chapter. As in Remark 6.1, for each $g,h\in G$ we shall choose any fixed representative of (g,h) in $\frac{1}{T}\mathbf{Z}$ and we shall continue to denote it (g,h); we assume that the representatives (g,h) and (h,g) agree, and that $(g,0)=0$. The notation z^g shall be interpreted accordingly.

In order to establish duality for generalized vertex operator algebras, we need some consequences of the Jacobi identity (6.12), namely, the "generalized commutator relation" and the "generalized associator relation," which extend the corresponding results for vertex operator algebras. Let $V=(V,Y,\mathbf{1},\omega,T,G,(\cdot,\cdot))$ be a generalized vertex operator algebra as in Chapter 6, and let $g,h\in G$ and $u\in V^g, v\in V^h$.

Recall from (6.11) that $z^{(g,h)}Y(u,z)v$ involves only integral powers of z. (Here we are beginning to use our convention that (g,h) is an element of $\mathbf{Q}$.) Accordingly, the operator

$$Y(u,z)z^g \text{ involves only integral powers of } z. \tag{7.3}$$

Remark 7.2: In view of (7.3), we shall take the liberty of substituting for z in $Y(u,z)z^g$ or $Y(u,z)z^{(g,h)}v$ expressions (such as $-z$) for which the corresponding substitutions in $Y(u,z)$, z^g and $z^{(g,h)}$ are not individually defined.

Our first consequence is the generalized commutator relation (7.5) below. First we multiply each side of the Jacobi identity (6.12) by $z_0^{(g,h)}z_1^g z_2^h$ on the right, in order to obtain operators whose expansions in powers of z_0, z_1 and z_2 involve only integral powers of these variables. Then we take Res_{z_0} and use (7.2) and (6.11) to obtain

$$\begin{aligned}
&(z_1-z_2)^{(g,h)}Y(u,z_1)Y(v,z_2)z_1^g z_2^h \\
&\quad -(z_2-z_1)^{(g,h)}Y(v,z_2)Y(u,z_1)z_1^g z_2^h \\
&= \mathrm{Res}_{z_0} z_2^{-1}\delta\left(\frac{z_1-z_0}{z_2}\right) Y(Y(u,z_0)z_0^{(g,h)}v,z_2)\left(\frac{z_1-z_0}{z_2}\right)^{-g} z_1^g z_2^h \\
&= \mathrm{Res}_{z_0} Y\left(\left(z_0^{-1}\delta\left(\frac{z_1-z_2}{z_0}\right) - z_0^{-1}\delta\left(\frac{z_2-z_1}{-z_0}\right)\right) Y(u,z_0)z_0^{(g,h)}v, z_2\right)\cdot \\
&\quad \cdot\left(\frac{z_1-z_0}{z_2}\right)^{-g} z_1^g z_2^h \\
&= \mathrm{Res}_{z_0} Y\left(\left(z_0^{-1}\delta\left(\frac{z_1-z_2}{z_0}\right) Y(u,z_1-z_2)(z_1-z_2)^{(g,h)}\right.\right. \\
&\quad \left.\left. -z_0^{-1}\delta\left(\frac{z_2-z_1}{-z_0}\right) Y(u,-z_2+z_1)(-z_2+z_1)^{(g,h)}\right) v, z_2\right)\cdot \\
&\quad \cdot\left(\frac{z_1-z_0}{z_2}\right)^{-g} z_1^g z_2^h.
\end{aligned}$$

But if $n \in \mathbf{Z}$ $(n \geq 0)$ is sufficiently large, say, $n \geq N$ (≥ 0),

$$\begin{aligned}
&\Big(Y(u,z_1-z_2)(z_1-z_2)^{(g,h)+n} \\
&\quad -Y(u,-z_2+z_1)(-z_2+z_1)^{(g,h)+n}\Big) v = 0,
\end{aligned} \tag{7.4}$$

so that we have the *generalized commutator relation*

$$(z_1-z_2)^{(g,h)}Y(u,z_1)Y(v,z_2)z_1^g z_2^h$$

$$
\begin{aligned}
&-(z_2 - z_1)^{(g,h)} Y(v, z_2) Y(u, z_1) z_1^g z_2^h \\
&= \sum_{n=0}^{N} (-1)^n Y\Big(\big(Y(u, z_1 - z_2)(z_1 - z_2)^{(g,h)+n} \\
&\qquad -Y(u, -z_2 + z_1)(-z_2 + z_1)^{(g,h)+n}\big)\, v, z_2\Big)\, z_1^{-n} z_2^{g+h} \binom{-g}{n},
\end{aligned} \tag{7.5}
$$

where $\binom{-g}{n}$ is the operator on V given by

$$
\binom{-g}{n} \cdot w = \binom{-(g,k)}{n} w \quad \text{for} \quad w \in V^k \ (k \in G). \tag{7.6}
$$

The two sides of (7.5) involve only integral powers of z_1 and z_2.

Remark 7.3: If V is a vertex operator algebra, i.e., $T = 1$ and $G = 0$, the identity (7.5) reduces to the commutation relation

$$
[Y(u, z_1), Y(v, z_2)] = Y((Y(u, z_1 - z_2) - Y(u, -z_2 + z_1))v, z_2) \tag{7.7}
$$

given in [FLM3], Remark 8.8.16.

To derive the second consequence, the generalized associator relation, we again multiply each of the three terms in the Jacobi identity by $z_0^{(g,h)} z_1^g z_2^h$ on the right, and we observe from (7.1) that

$$
\begin{aligned}
&z_2^{-1} \delta\left(\frac{z_1 - z_0}{z_2}\right) Y(Y(u, z_0)v, z_2) \left(\frac{z_1 - z_0}{z_2}\right)^{-g} z_1^g z_0^{(g,h)} z_2^h \\
&= z_1^{-1} \delta\left(\frac{z_2 + z_0}{z_1}\right) Y(Y(u, z_0)v, z_2)(z_2 + z_0)^g z_0^{(g,h)} z_2^h.
\end{aligned} \tag{7.8}
$$

Next we see from (7.1) and (5.5) that

$$
\begin{aligned}
&z_0^{-1} \left(\frac{z_1 - z_2}{z_0}\right)^{(g,h)} \delta\left(\frac{z_1 - z_2}{z_0}\right) Y(u, z_1) Y(v, z_2) z_0^{(g,h)} z_1^g z_2^h \\
&= z_1^{-1} \left(\frac{z_0 + z_2}{z_1}\right)^{-(g,h)} \delta\left(\frac{z_0 + z_2}{z_1}\right) Y(u, z_1) Y(v, z_2) z_0^{(g,h)} z_1^g z_2^h \\
&= z_1^{-1} \delta\left(\frac{z_0 + z_2}{z_1}\right) Y(u, z_0 + z_2) Y(v, z_2) z_0^{(g,h)} (z_0 + z_2)^g z_2^h,
\end{aligned}
$$

since these expressions involve only integral powers of z_1. Now we take Res_{z_1} and use (7.1) and (7.2) to see that

$$\begin{aligned}
&Y(Y(u,z_0)v,z_2)(z_2+z_0)^g z_0^{(g,h)} z_2^h \\
&\quad -Y(u,z_0+z_2)Y(v,z_2)(z_0+z_2)^g z_0^{(g,h)} z_2^h \\
&\quad = -\mathrm{Res}_{z_1} z_0^{-1}\left(\frac{z_2-z_1}{z_0}\right)^{(g,h)} \delta\left(\frac{z_2-z_1}{-z_0}\right) Y(v,z_2)Y(u,z_1) z_0^{(g,h)} z_1^g z_2^h \\
&\quad = \mathrm{Res}_{z_1} Y(v,z_2)\left\{ z_2^{-1}\delta\left(\frac{z_1-z_0}{z_2}\right) Y(u,z_1) z_1^g \right. \\
(7.9)\quad &\quad \left. -z_0^{-1}\delta\left(\frac{z_1-z_2}{z_0}\right) Y(u,z_1) z_1^g \right\} (z_2-z_1)^{(g,h)} z_2^h \\
&\quad = \mathrm{Res}_{z_1} Y(v,z_2)\left\{ z_1^{-1}\delta\left(\frac{z_2+z_0}{z_1}\right) Y(u,z_1) z_1^g \right. \\
&\quad \left. -z_1^{-1}\delta\left(\frac{z_0+z_2}{z_1}\right) Y(u,z_1) z_1^g \right\} (z_2-z_1)^{(g,h)} z_2^h \\
&\quad = \mathrm{Res}_{z_1} Y(v,z_2)\left\{ z_1^{-1}\delta\left(\frac{z_2+z_0}{z_1}\right) Y(u,z_2+z_0)(z_2+z_0)^g \right. \\
&\quad \left. -z_1^{-1}\delta\left(\frac{z_0+z_2}{z_1}\right) Y(u,z_0+z_2)(z_0+z_2)^g \right\} (z_2-z_1)^{(g,h)} z_2^h .
\end{aligned}$$

For $w \in V$, if $n \in \mathbf{Z}$ $(n \geq 0)$ is large enough, say, $n \geq N$ (≥ 0), we have

$$\begin{aligned}
(7.10)\qquad &\Big(Y(u,z_2+z_0)(z_2+z_0)^{g+n} \\
&\quad -Y(u,z_0+z_2)(z_0+z_2)^{g+n}\Big) w = 0,
\end{aligned}$$

and we conclude that

$$\begin{aligned}
&\Big(Y(Y(u,z_0)v,z_2)(z_2+z_0)^g z_0^{(g,h)} z_2^h \\
&\quad -Y(u,z_0+z_2)Y(v,z_2)(z_0+z_2)^g z_0^{(g,h)} z_2^h\Big) w \\
(7.11)\qquad &\quad = \sum_{n=0}^{N} (-1)^n Y(v,z_2)\Big(Y(u,z_2+z_0)(z_2+z_0)^{g+n} \\
&\quad -Y(u,z_0+z_2)(z_0+z_2)^{g+n}\Big) z_2^{(g,h)+h-n} \binom{(g,h)}{n} w.
\end{aligned}$$

We call this the *generalized associator relation.* The two sides involve only integral powers of z_0 and z_2.

Remark 7.4: If V is a vertex operator algebra, (7.11) becomes

$$\begin{aligned}
(7.12)\qquad &Y(Y(u,z_0)v,z_2) - Y(u,z_0+z_2)Y(v,z_2) \\
&\quad = Y(v,z_2)(Y(u,z_2+z_0) - Y(u,z_0+z_2))
\end{aligned}$$

([FLM3], formula (8.10.61)). We call this formula the *associator relation* because the left-hand side can be thought of as an associator, in the spirit of the associativity relation [FLM3], Proposition 8.10.5.

The identities (7.5) and (7.11) will be used to establish generalized commutativity and generalized associativity, respectively, for generalized vertex operator algebras. In order to formulate these duality properties, we also need the following discussion on expansions of rational functions such as $(z_1 \pm z_2)^n$ for $n \in \mathbf{Z}$ (cf. [FLM3], Sections 8.1, 8.10 and [FHL]).

Let S denote the set of nonzero linear polynomials in two variables, say, z_1 and z_2 :

$$S = \{c_1 z_1 + c_2 z_2 | c_1, c_2 \in \mathbf{C} \text{ not both zero}\} \subset \mathbf{C}[z_1, z_2]. \tag{7.13}$$

Consider the subring $\mathbf{C}[z_1, z_2]_S$ of the field of rational functions $\mathbf{C}(z_1, z_2)$ obtained by inverting the products of (zero or more) elements of S. Let the pair (i_1, i_2) be either (1,2) or (2,1). We define the linear and multiplicative map

$$\iota_{i_1 i_2} : \mathbf{C}[z_1, z_2]_S \to \mathbf{C}[[z_1, z_1^{-1}, z_2, z_2^{-1}]] \tag{7.14}$$

so that $\iota_{i_1 i_2}$ is the identity on $\mathbf{C}[z_1, z_1^{-1}, z_2, z_2^{-1}]$ and so that $\iota_{i_1 i_2}((c_1 z_1 + c_2 z_2)^{-1})$ is the expansion of $(c_1 z_1 + c_2 z_2)^{-1}$ in nonnegative integral powers of z_{i_2}.

The following formal algebraic analogue of the Cauchy residue theorem for suitably special meromorphic functions, which follows immediately from (7.1) (with $r = 0$) and (7.2), will be used below to establish the Jacobi identity from duality:

Proposition 7.5 *[FHL] Consider a rational function of the form*

$$f(z_0, z_1, z_2) = \frac{g(z_0, z_1, z_2)}{z_0^{r_0} z_1^{r_1} z_2^{r_2}} \tag{7.15}$$

where g is a polynomial and $r_i \in \mathbf{Z}$. Then

$$z_1^{-1}\delta\left(\frac{z_2 + z_0}{z_1}\right)\iota_{20}(f|_{z_1 = z_0 + z_2}) = z_2^{-1}\delta\left(\frac{z_1 - z_0}{z_2}\right)\iota_{10}(f|_{z_2 = z_1 - z_0}) \tag{7.16}$$

and

$$\begin{aligned} &z_0^{-1}\delta\left(\frac{z_1 - z_2}{z_0}\right)\iota_{12}(f|_{z_0 = z_1 - z_2}) - z_0^{-1}\delta\left(\frac{z_2 - z_1}{-z_0}\right)\iota_{21}(f|_{z_0 = z_1 - z_2}) \\ &= z_2^{-1}\delta\left(\frac{z_1 - z_0}{z_2}\right)\iota_{10}(f|_{z_2 = z_1 - z_0}). \quad \square \end{aligned} \tag{7.17}$$

We now apply this formalism to generalized vertex operator algebras by considering "matrix coefficients" of products of vertex operators. Set

$$V' = \coprod_{n \in \mathbf{Q}} V_n^*, \tag{7.18}$$

the direct sum of the dual spaces of the homogeneous subspaces V_n of V – the space of linear functionals on V vanishing on all but finitely many V_n. Denote by $\langle \cdot, \cdot \rangle$ the natural pairing between V' and V. For some purposes below, we shall be restricting our attention to subsets or subspaces A of V' which are *dense* in the sense that for $v \in V$, $\langle A, v \rangle = 0$ implies $v = 0$.

For $g, h \in G, v_1 \in V^g, v \in V^h$ and $v' \in V'$,

$$\langle v', Y(v_1, z_1)v \rangle z_1^{(g,h)} \in \mathbf{C}[z_1, z_1^{-1}] \tag{7.19}$$

by (6.11) and Remark 6.3. We have:

Proposition 7.6 *(a)* **(generalized rationality of products)** *For $g_1, g_2, h \in G$ and $v_1 \in V^{g_1}, v_2 \in V^{g_2}, v \in V^h, v' \in V'$, the formal series*

$$\langle v', Y(v_1, z_1)Y(v_2, z_2)v \rangle (z_1 - z_2)^{(g_1,g_2)} z_1^{(g_1,h)} z_2^{(g_2,h)},$$

which is well defined and which involves only integral powers of z_1 and z_2, lies in the image of the map ι_{12} :

$$\begin{aligned} &\langle v', Y(v_1, z_1)Y(v_2, z_2)v \rangle (z_1 - z_2)^{(g_1,g_2)} z_1^{(g_1,h)} z_2^{(g_2,h)} \\ &\quad = \iota_{12} f(z_1, z_2), \end{aligned} \tag{7.20}$$

where the (uniquely determined) element $f \in \mathbf{C}[z_1, z_2]_S$ is of the form

$$f(z_1, z_2) = g(z_1, z_2)/z_1^{r_1} z_2^{r_2} (z_1 - z_2)^t \tag{7.21}$$

for some $g \in \mathbf{C}[z_1, z_2]$ and $r_1, r_2, t \in \mathbf{Z}$. The integer t depends only on v_1 and v_2; it is independent of v and v'.

(b) **(generalized commutativity)** *We also have*

$$\begin{aligned} &\langle v', Y(v_2, z_2)Y(v_1, z_1)v \rangle (z_2 - z_1)^{(g_1,g_2)} z_1^{(g_1,h)} z_2^{(g_2,h)} \\ &\quad = \iota_{21} f(z_1, z_2) \end{aligned} \tag{7.22}$$

(in particular, the left-hand side is well defined), that is,

$$\begin{aligned} &\text{“}Y(v_1, z_1)Y(v_2, z_2)(z_1 - z_2)^{(g_1,g_2)} z_1^{g_1} z_2^{g_2} \quad \text{agrees with} \\ &\qquad Y(v_2, z_2)Y(v_1, z_1)(z_2 - z_1)^{(g_1,g_2)} z_1^{g_1} z_2^{g_2} \end{aligned} \tag{7.23}$$

as operator-valued rational functions."

Proof By (7.4) and (7.5), if t is a large enough nonnegative integer,

$$\begin{aligned}(7.24)\qquad & Y(v_1,z_1)Y(v_2,z_2)(z_1-z_2)^{(g_1,g_2)}z_1^{g_1}z_2^{g_2}(z_1-z_2)^t \\ & = Y(v_2,z_2)Y(v_1,z_1)(z_2-z_1)^{(g_1,g_2)}z_1^{g_1}z_2^{g_2}(z_1-z_2)^t,\end{aligned}$$

and

$$\begin{aligned}(7.25)\qquad & \langle v', Y(v_1,z_1)Y(v_2,z_2)v\rangle(z_1-z_2)^{(g_1,g_2)}z_1^{(g_1,h)}z_2^{(g_2,h)}(z_1-z_2)^t \\ & = \langle v', Y(v_2,z_2)Y(v_1,z_1)v\rangle(z_2-z_1)^{(g_1,g_2)}z_1^{(g_1,h)}z_2^{(g_2,h)}(z_1-z_2)^t\end{aligned}$$

involves only integral powers of z_1 and z_2. But the left-hand side of (7.25) involves only finitely many positive powers z_1, by Remark 6.3, and the right-hand side involves only finitely many negative powers of z_1, by (6.8). Thus each side of (7.25) involves only finitely many powers of z_1. Similarly, each side of (7.25) involves only finitely many powers of z_2, so that each side of (7.25) is of the form $h(z_1,z_2) \in \mathbf{C}[z_1,z_1^{-1},z_2,z_2^{-1}]$. Then

$$f(z_1,z_2) = h(z_1,z_2)/(z_1-z_2)^t$$

satisfies the desired conditions. In fact, the left-hand side of (7.20) involves only finitely many negative powers of z_2 and so can be multiplied by $(z_1-z_2)^{-t}$, and analogously, the left-hand side of (7.22) involves only finitely many negative powers of z_1 and so can be multiplied by $(-z_2+z_1)^{-t}$. □

Remark 7.7: Note that the assertion of Proposition 7.6 (even if v' is taken to range only through a dense subset or subspace of V') implies (7.24). Thus we have the important observation that formula (7.24) is equivalent to the assertion of Proposition 7.6. In particular, the assertion of Proposition 7.6 with v' restricted to a dense subset implies Proposition 7.6 in full.

An analogous argument using (7.11) gives:

Proposition 7.8 *(a)* **(generalized rationality of iterates)** *For $g_1, g_2, h \in G$ and $v_1 \in V^{g_1}, v_2 \in V^{g_2}, v \in V^h, v' \in V'$, the formal series*

$$\langle v', Y(Y(v_1,z_0)v_2,z_2)v\rangle(z_2+z_0)^{(g_1,h)}z_2^{(g_2,h)}z_0^{(g_1,g_2)},$$

which is well defined and which involves only integral powers z_0 and z_2, lies in the image of the map ι_{20} :

$$\begin{aligned}(7.26)\qquad & \langle v', Y(Y(v_1,z_0)v_2,z_2)v\rangle(z_2+z_0)^{(g_1,h)}z_2^{(g_2,h)}z_0^{(g_1,g_2)} \\ & = \iota_{20}h(z_0,z_2),\end{aligned}$$

where the (uniquely determined) element $h \in \mathbf{C}[z_0, z_2]_S$ *is of the form*

$$h(z_0, z_2) = k(z_0, z_2)/z_0^{j_0} z_2^{j_2} (z_0 + z_2)^l \tag{7.27}$$

for some $k(z_0, z_2) \in \mathbf{C}[z_0, z_2]$ *and* $j_0, j_2, l \in \mathbf{Z}$. *The integer* l *depends only on* v_1 *and* v; *it is independent of* v_2 *and* v'.
(b) We also have

$$\begin{aligned} &\langle v', Y(v_1, z_0 + z_2) Y(v_2, z_2) v\rangle (z_0 + z_2)^{(g_1,h)} z_2^{(g_2,h)} z_0^{(g_1,g_2)} \\ &\quad = \iota_{02} h(z_0, z_2) \end{aligned} \tag{7.28}$$

(in particular, the left-hand side is well defined), that is,

$$\begin{aligned} &\text{“}Y(Y(v_1, z_0)v_2, z_2)(z_2 + z_0)^{g_1} z_2^{g_2} z_0^{(g_1,g_2)} \text{ agrees with} \\ &\quad Y(v_1, z_0 + z_2) Y(v_2, z_2)(z_0 + z_2)^{g_1} z_2^{g_2} z_0^{(g_1,g_2)} \end{aligned} \tag{7.29}$$

as operator-valued rational functions.” □

It is clear that for the rational function $f(z_1, z_2)$ of (7.21),

$$\iota_{02} f(z_0 + z_2, z_2) = (\iota_{12} f(z_1, z_2))|_{z_1 = z_0 + z_2}, \tag{7.30}$$

so that from (7.20) and (7.28),

$$h(z_0, z_2) = f(z_0 + z_2, z_2). \tag{7.31}$$

Thus the last two propositions give:

Proposition 7.9 (generalized associativity) *We have:*

$$\begin{aligned} &\iota_{12}^{-1} \left(\langle v', Y(v_1, z_1) Y(v_2, z_2) v\rangle (z_1 - z_2)^{(g_1,g_2)} z_1^{(g_1,h)} z_2^{(g_2,h)} \right) \\ &= \iota_{20}^{-1} \left(\langle v', Y(Y(v_1, z_0)v_2, z_2) v\rangle (z_2 + z_0)^{(g_1,h)} z_2^{(g_2,h)} z_0^{(g_1,g_2)} \right) \big|_{z_0 = z_1 - z_2}. \end{aligned} \tag{7.32}$$

That is,

$$\begin{aligned} &\text{“}Y(v_1, z_1) Y(v_2, z_2)(z_1 - z_2)^{(g_1,g_2)} z_1^{g_1} z_2^{g_2} \text{ agrees with} \\ &\quad Y(Y(v_1, z_1 - z_2)v_2, z_2)(z_1 - z_2)^{(g_1,g_2)} z_1^{g_1} z_2^{g_2} \end{aligned} \tag{7.33}$$

as operator-valued rational functions, where the right-hand expression is to be expanded as a Laurent series in $z_1 - z_2$.*”* □

Corollary 7.10 *(a) The formal series obtained by taking matrix coefficients of the two expressions in (7.23) converge to a common rational function in the (disjoint) domains*

(7.34) $$|z_1| > |z_2| > 0 \text{ and } |z_2| > |z_1| > 0,$$

respectively.
(b) The formal series obtained by taking matrix coefficients of the two expressions in (7.33) converge to a common rational function in the domains

(7.35) $$|z_1| > |z_2| > 0 \text{ and } |z_2| > |z_1 - z_2| > 0,$$

respectively, and in the common domain

(7.36) $$|z_1| > |z_2| > |z_1 - z_2| > 0,$$

these two series converge to the common function. □

Conversely, from Proposition 7.5 and (7.8) we have:

Proposition 7.11 *The generalized Jacobi identity follows from the generalized rationality of products and iterates, and generalized commutativity and associativity. In particular, in the definition of the notion of generalized vertex operator algebra, the Jacobi identity can be replaced by these properties. Moreover, this equivalence still holds if in the statements of generalized rationality, commutativity and associativity, the vector v' is restricted to range through a fixed dense subset or subspace of V'.* □

To discuss representations of braid groups in the next chapter, we extend generalized rationality and commutativity to several variables. For this, we generalize the $\iota_{i_1 i_2}$ notation to n variables, say, $z_1, ..., z_n$, following [FLM3], Section 8.10 and [FHL]. The set S of (7.13) is now the set of nonzero homogeneous polynomials in $z_1, ..., z_n$. For a permutation $(i_1 \cdots i_n)$ of $(1 \cdots n)$, the linear and multiplicative map

(7.37) $$\iota_{i_1 \cdots i_n} : \mathbf{C}[z_1, ..., z_n]_S \to \mathbf{C}[[z_1, z_1^{-1} ..., z_n, z_n^{-1}]]$$

is defined by analogy with (7.14), where we expand successively in nonnegative integral powers of z_{i_n}, then $z_{i_{n-1}}$, and so on (see [FLM3], Section 8.10 or [FHL] for a formal inductive definition). This map is injective.

Proposition 7.12 (generalized rationality and commutativity) *For $g_1, ..., g_n, h \in G$ and $v_1 \in V^{g_1}, ..., v_n \in V^{g_n}, v \in V^h, v' \in V'$ and any permutation $(i_1 \cdots i_n)$ of $(1 \cdots n)$, the formal series*

$$\langle v', Y(v_{i_1}, z_{i_1}) \cdots Y(v_{i_n}, z_{i_n})v\rangle \prod_{j<k} (z_{i_j} - z_{i_k})^{(g_{i_j}, g_{i_k})} \prod_{l=1}^{n} z_l^{(g_l, h)},$$

which is well defined and which involves only integral powers of each z_i, lies in the image of the map $\iota_{i_1 \cdots i_n}$:

$$\begin{aligned} &\langle v', Y(v_{i_1}, z_{i_1}) \cdots Y(v_{i_n}, z_{i_n})v\rangle \prod_{j<k} (z_{i_j} - z_{i_k})^{(g_{i_j}, g_{i_k})} \prod_{l=1}^{n} z_l^{(g_l, h)} \\ &\quad = \iota_{i_1 \cdots i_n} f(z_1, ..., z_n), \end{aligned} \tag{7.38}$$

where the (uniquely determined) element $f \in \mathbf{C}[z_1, ..., z_n]_S$ is independent of the permutation and is of the form

$$f(z_1, ..., z_n) = g(z_1, ..., z_n) / \prod_{i=1}^{n} z_i^{r_i} \prod_{j<k} (z_j - z_k)^{t_{jk}} \tag{7.39}$$

for some $g \in \mathbf{C}[z_1, ..., z_n]$ and $r_i, t_{jk} \in \mathbf{Z}$. The integer t_{jk} depends only on v_j and v_k; it is independent of v_l $(l \neq j, k)$, v and v'. In particular,

$$\begin{aligned} &\text{“}Y(v_1, z_1) \cdots Y(v_n, z_n) \prod_{j<k} (z_j - z_k)^{(g_j, g_k)} \prod_{l=1}^{n} z_l^{g_l} \text{ agrees with} \\ &Y(v_{i_1}, z_{i_1}) \cdots Y(v_{i_n}, z_{i_n}) \prod_{j<k} (z_{i_j} - z_{i_k})^{(g_{i_j}, g_{i_k})} \prod_{l=1}^{n} z_l^{g_l} \end{aligned} \tag{7.40}$$

as operator-valued rational functions.”

Proof This proof generalizes that of Proposition 7.6. First we show that the left-hand side of (7.38) is well defined. Without loss of generality, we assume here that $(i_1 \cdots i_n) = (1 \cdots n)$, and we show by induction on n that

$$Y(v_1, z_1) \cdots Y(v_n, z_n)v \prod_{j<k} (z_j - z_k)^{(g_j, g_k)} \prod_{l=1}^{n} z_l^{(g_l, h)} \tag{7.41}$$

exists and involves only integral powers of each z_i. This holds for $n = 1$ by (7.3) (and for $n = 2$ by Proposition 7.6). But the coefficient of each power of z_n in (7.41), which involves only integral powers of z_n, is a finite linear combination, with Laurent monomial coefficients, of analogous expressions for $n-1$, with $v \in V^{g_n+h}$; here we are using the fact that

$$(g_l, g_n + h) \equiv (g_l, g_n) + (g_l, h) \mod 2\mathbf{Z}$$

(and hence mod $\mathbf{Z}$). The induction result follows.

Next, by (7.24) (or by Proposition 7.6; recall Remark 7.7), for $j \neq k$, there exists a nonnegative integer t_{jk} such that

$$\begin{aligned} &Y(v_j, z_j)Y(v_k, z_k)(z_j - z_k)^{(g_j,g_k)} z_j^{g_j} z_k^{g_k} (z_j - z_k)^{t_{jk}} \\ &= Y(v_k, z_k)Y(v_j, z_j)(z_k - z_j)^{(g_j,g_k)} z_j^{g_j} z_k^{g_k} (z_j - z_k)^{t_{jk}}. \end{aligned} \tag{7.42}$$

Thus

$$\begin{aligned} &Y(v_1, z_1) \cdots Y(v_n, z_n) \prod_{j<k} (z_j - z_k)^{(g_j,g_k)} \prod_{l=1}^{n} z_l^{g_l} \prod_{j<k} (z_j - z_k)^{t_{jk}} \\ &= Y(v_{i_1}, z_{i_1}) \cdots Y(v_{i_n}, z_{i_n}) \cdot \\ &\cdot \prod_{j<k} (z_{i_j} - z_{i_k})^{(g_{i_j},g_{i_k})} \prod_{l=1}^{n} z_l^{g_l} \prod_{j<k} (z_j - z_k)^{t_{jk}}. \end{aligned} \tag{7.43}$$

But

$$\begin{aligned} &\langle v', Y(v_{i_1}, z_{i_1}) \cdots Y(v_{i_n}, z_{i_n}) v \rangle \cdot \\ &\cdot \prod_{j<k} (z_{i_j} - z_{i_k})^{(g_{i_j},g_{i_k})} \prod_{l=1}^{n} z_l^{(g_l,h)} \prod_{j<k} (z_j - z_k)^{t_{jk}} \end{aligned} \tag{7.44}$$

involves only finitely many positive (integral) powers of z_{i_1}, by Remark 6.3, and only finitely many negative powers of z_{i_n}, by (6.8), and hence it involves only finitely many powers of each z_i, since the permutation is arbitrary. Thus (7.44) is of the form

$$h(z_1, \cdots, z_n) \in \mathbf{C}[z_1, z_1^{-1}, \cdots, z_n, z_n^{-1}].$$

The rational function

$$f(z_1, ..., z_n) = h(z_1, ..., z_n) / \prod_{j<k} (z_j - z_k)^{t_{jk}}$$

satisfies the desired conditions. In fact, a simpler version of the inductive argument at the beginning of the proof shows that the formal Laurent series in (7.41) can be multiplied by the formal Laurent series $\prod_{j<k}(z_j - z_k)^{-t_{jk}}$. This establishes (7.38) for $(i_1 \cdots i_n) = (1 \cdots n)$, and the analogous argument proves (7.38) for an arbitrary permutation. □

Remark 7.13: In view of Remark 7.7, this proof, which uses Remark 6.3 and hence formula (6.21) (the $L(0)$-bracket property), actually shows: Suppose that we assume all the axioms for a generalized vertex operator algebra except for the Jacobi identity (6.12), and that we assume (6.21). Then generalized rationality and commutativity for two vertex operators (the assertion of Proposition 7.6), even in the weakened form mentioned in Remark 7.7, implies generalized rationality and commutativity for an arbitrary number of vertex operators (the assertion of Proposition 7.12).

Remark 7.14: In case $v = \mathbf{1}$ in Proposition 7.12, formula (6.10) and the fact that $(g,0) = (0,g) = 0$ for $g \in G$ show that (7.38) takes the form

$$\begin{aligned}(7.45)\qquad &\langle v', Y(v_{i_1}, z_{i_1}) \cdots Y(v_{i_n}, z_{i_n})\mathbf{1}\rangle \prod_{j<k} (z_{i_j} - z_{i_k})^{(g_{i_j}, g_{i_k})} \\ &= \iota_{i_1 \cdots i_n} f(z_1, ..., z_n),\end{aligned}$$

where f has no pole at $z_l = 0$ $(l = 1, ..., n)$:

$$(7.46)\qquad f(z_1, ..., z_n) = g(z_1, ..., z_n)/\prod_{j<k}(z_j - z_k)^{t_{jk}}.$$

Corollary 7.15 *The formal series obtained by taking matrix coefficients of the two expressions in (7.40) converge to a common rational function in the domains*

$$(7.47)\qquad |z_1| > \cdots > |z_n| > 0 \quad \text{and} \quad |z_{i_1}| > \cdots > |z_{i_n}| > 0,$$

respectively. □

Next, we show essentially that generalized rationality and commutativity for two vertex operators implies generalized associativity and hence the Jacobi identity. We begin by assuming the axioms for a generalized vertex operator algebra except for the Jacobi identity, and we also assume the validity of Proposition 7.12 (generalized rationality and commutativity) for the special case of three operators applied to $v = \mathbf{1}$, together with the $L(-1)$-bracket formula (6.20).

First note that (6.26)-(6.28) hold. Let $g_i \in G$ and $v_i \in V^{g_i}$ for

$i = 1, 2, 3$, and let $v' \in V'$. Then

$$\begin{aligned}
&\langle v', Y(v_3, z_3)Y(Y(v_1, z_0)v_2, z_2)\mathbf{1}\rangle \\
&\quad = \langle v', Y(v_3, z_3)e^{z_2 L(-1)}Y(v_1, z_0)v_2\rangle \\
&\quad = \langle v', Y(v_3, z_3)Y(v_1, z_0 + z_2)e^{z_2 L(-1)}v_2\rangle \\
&\quad = \langle v', Y(v_3, z_3)Y(v_1, z_0 + z_2)Y(v_2, z_2)\mathbf{1}\rangle,
\end{aligned} \tag{7.48}$$

where of course $Y(v_1, z_0 + z_2)$ is to be expanded in nonnegative integral powers of z_2. Multiplying either side of (7.48) by $z_0^{(g_1,g_2)} z_3^{(g_1,g_3)} z_3^{(g_2,g_3)}$ turns it into a series involving only integral powers of z_0, z_2 and z_3. (Recall the bilinearity of $(\cdot,\cdot)$ mod $2\mathbf{Z}$ and hence mod $\mathbf{Z}$.) We would like essentially to move $Y(v_3, z_3)$ to the right in the first and last expressions in (7.48) and then set $z_3 = 0$.

By assumption and Remark 7.14, we may write

$$\begin{aligned}
&\langle v', Y(v_1, z_1)Y(v_2, z_2)Y(v_3, z_3)\mathbf{1}\rangle \prod_{j<k}(z_j - z_k)^{(g_j,g_k)} \\
&\quad = \iota_{123} f(z_1, z_2, z_3),
\end{aligned} \tag{7.49}$$

where

$$f(z_1, z_2, z_3) = g(z_1, z_2, z_3)/(z_1 - z_2)^p (z_1 - z_3)^q (z_2 - z_3)^r, \tag{7.50}$$

f has the permutation-independence property, and each of p, q, r depends only on a certain pair of vectors.

Regarding operations such as ι_{23} for expressions in three variables as acting on each expansion coefficient with respect to z_0, we see, using Remark 7.14 for $n = 2$ and the bilinearity of $(\cdot,\cdot)$ modulo $2\mathbf{Z}$, that as formal series in z_0 with coefficients which are rational functions of z_2 and z_3, for $m \in \mathbf{Z}$,

$$\begin{aligned}
&\iota_{23}^{-1}(\langle v', Y(Y(v_1, z_0)v_2, z_2)Y(v_3, z_3)\mathbf{1}\rangle \cdot \\
&\quad \cdot z_0^{(g_1,g_2)}(z_2 - z_3)^{(g_1,g_3)+(g_2,g_3)+m}) \\
&\quad = (-1)^m \iota_{32}^{-1}(\langle v', Y(v_3, z_3)Y(Y(v_1, z_0)v_2, z_2)\mathbf{1}\rangle \cdot \\
&\quad \cdot z_0^{(g_1,g_2)}(z_3 - z_2)^{(g_1,g_3)+(g_2,g_3)+m}).
\end{aligned} \tag{7.51}$$

(Note that the bilinearity of $(\cdot,\cdot)$ mod $2\mathbf{Z}$ and not just mod $\mathbf{Z}$ is really needed here.) Thus from (7.48),

$$\iota_{23}^{-1}(\langle v', Y(Y(v_1, z_0)v_2, z_2)Y(v_3, z_3)\mathbf{1}\rangle \cdot$$

$$\cdot z_0^{(g_1,g_2)}(z_2 - z_3)^{(g_2,g_3)}(z_2 - z_3 + z_0)^{(g_1,g_3)})$$
$$= \iota_{32}^{-1}(\langle v', Y(v_3, z_3)Y(Y(v_1, z_0)v_2, z_2)\mathbf{1}\rangle \cdot$$
$$\cdot z_0^{(g_1,g_2)}(z_3 - z_2)^{(g_2,g_3)}(z_3 - z_2 - z_0)^{(g_1,g_3)})$$
$$= \iota_{32}^{-1}(\langle v', Y(v_3, z_3)Y(v_1, z_0 + z_2)Y(v_2, z_2)\mathbf{1}\rangle \cdot$$
$$\cdot z_0^{(g_1,g_2)}(z_3 - z_2)^{(g_2,g_3)}(z_3 - z_0 - z_2)^{(g_1,g_3)})$$
$$= \iota_{32}^{-1}(\iota_{312}f(z_1, z_2, z_3)|_{z_1=z_0+z_2})$$
$$= \iota_{32}^{-1}(g(z_1, z_2, z_3)\iota_{12}((z_1 - z_2)^{-p}) \cdot$$
$$\cdot \iota_{31}((z_1 - z_3)^{-q})\iota_{32}((z_2 - z_3)^{-r})|_{z_1=z_0+z_2})$$
$$= \iota_{32}^{-1}(g(z_0 + z_2, z_2, z_3)z_0^{-p} \cdot$$
$$\cdot \iota_{320}((z_0 + z_2 - z_3)^{-q})\iota_{32}((z_2 - z_3)^{-r}))$$
$$= g(z_0 + z_2, z_2, z_3)z_0^{-p}\iota_{40}((z_0 + z_4)^{-q})z_4^{-r}|_{z_4=z_2-z_3}, \tag{7.52}$$

so that

$$\langle v', Y(Y(v_1, z_0)v_2, z_2)Y(v_3, z_3)\mathbf{1}\rangle \cdot$$
$$\cdot z_0^{(g_1,g_2)}(z_2 - z_3)^{(g_2,g_3)}(z_2 - z_3 + z_0)^{(g_1,g_3)}$$
$$= \iota_{230}(g(z_0 + z_2, z_2, z_3)z_0^{-p}(z_0 + z_2 - z_3)^{-q}(z_2 - z_3)^{-r}) \tag{7.53}$$
$$= \iota_{230}f(z_0 + z_2, z_2, z_3).$$

On the other hand,

$$\langle v', Y(v_1, z_0 + z_2)Y(v_2, z_2)Y(v_3, z_3)\mathbf{1}\rangle \cdot$$
$$\cdot (z_0 + z_2 - z_3)^{(g_1,g_3)}(z_2 - z_3)^{(g_2,g_3)}z_0^{(g_1,g_2)}$$
$$= (\iota_{123}f(z_1, z_2, z_3))|_{z_1=z_0+z_2} \tag{7.54}$$
$$= \iota_{023}f(z_0 + z_2, z_2, z_3).$$

Thus the left-hand side of (7.53) and (7.54) are two different expansions of the same rational function. We may set $z_3 = 0$ in both of these expansions and in the common rational function, and we find that

$$\begin{aligned}&\langle v', Y(Y(v_1, z_0)v_2, z_2)v_3\rangle(z_2 + z_0)^{(g_1,g_3)}z_2^{(g_2,g_3)}z_0^{(g_1,g_2)}\\ &= \iota_{20}f(z_0 + z_2, z_2, 0)\end{aligned} \tag{7.55}$$

$$\begin{aligned}&\langle v', Y(v_1, z_0 + z_2)Y(v_2, z_2)v_3\rangle(z_0 + z_2)^{(g_1,g_3)}z_2^{(g_2,g_3)}z_0^{(g_1,g_2)}\\ &= \iota_{02}f(z_0 + z_2, z_2, 0).\end{aligned} \tag{7.56}$$

This gives us the assertions of Proposition 7.8 and hence 7.9. Recalling Proposition 7.11 and Remark 7.13, we see that we have established:

Proposition 7.16 *Generalized rationality of iterates and generalized associativity follow from generalized rationality and commutativity (for products of two vertex operators), the $L(-1)$-bracket formula (6.20), the $L(0)$-bracket formula (6.21) and the axioms for generalized vertex operator algebras except for the Jacobi identity. In particular, the Jacobi identity may be replaced by generalized rationality and commutativity, (6.20) and (6.21) in the axioms.* □

Remark 7.17: The argument has shown that we may replace the hypotheses in Proposition 7.16 by generalized rationality and commutativity for products of three vertex operators applied to $\mathbf{1}$ (the case $n = 3$, $v = \mathbf{1}$ in Proposition 7.12), (6.20) and the axioms except for the Jacobi identity. In particular, the Jacobi identity may be replaced by such generalized commutativity and (6.20) in the axioms.

Remark 7.18: Of course, Proposition 7.16 or Remark 7.17 suggests an alternate approach for proving the (generalized) Jacobi identity (cf. [FLM3, Appendix], [Go]). See also Chapter 9 below.

Remark 7.19: As in [FHL], we observe that the entire discussion of duality for algebras in this chapter, through Remark 7.13 (and also including Corollary 7.15), also applies, with the obvious modifications, to modules, even in the greater generality expressed in Remark 6.15.

Chapter 8

Monodromy representations of braid groups

Here we use the results in the last chapter on generalized rationality and commutativity to show that when the formal variables are specialized to complex variables, the monodromy (multiple-valuedness) associated with products of several vertex operators naturally produces one-dimensional braid group representations. In fact, suitable matrix coefficients of such products are the formal series expansions in appropriate domains, depending on the order of the vertex operators, of certain algebraic functions. Paying careful attention to the infinite-dimensionality of certain spaces, we adapt some of the setting used in [TK] to our situation in order to give a precise, conceptual description of the resulting braid group representations. We observe that these representations factor through certain Hecke algebras. The nonzero complex scalars according to which the generators of the braid group act on the constructed one-dimensional modules will be conceptually interpreted in Chapter 12 in terms of certain third cohomology. As is the case for the duality results in Chapter 7, the results in this chapter will be transferred in later chapters to the greater generalities of generalized vertex algebras and abelian intertwining algebras. As in Chapter 7, the reader should observe how the considerations in this chapter simplify in the "single-valued" case of generalized vertex operator algebras for which the form $(\cdot,\cdot)$ on G takes values in $\mathbf{Z}/2\mathbf{Z}$, in particular, for vertex operator superalgebras.

In order to study monodromy properties of the "multi-point correlation functions"

$$\langle v', Y(v_1, z_1) \cdots Y(v_n, z_n) v \rangle, \tag{8.1}$$

we discuss algebraic functions such as $(z_1 - z_2)^r$ for $r \in \mathbf{Q}$. In this chapter, the z_i will be understood as complex variables rather than formal variables. First we consider single-valued functions on restricted domains. For $z \in \mathbf{C}\backslash\{z \in \mathbf{R} | z \leq 0\}$ and $r \in \mathbf{Q}$, z^r will signify

$$z^r = e^{r(\log|z| + i\arg z)}, \quad -\pi < \arg z < \pi. \tag{8.2}$$

In particular, $(z_1 - z_2)^r$ is defined for $z_1 \not\leq z_2$ (i.e., $z_1 - z_2$ not a non-positive real number). Then for $z \notin \mathbf{R}$,

$$(-z)^r = e^{-\epsilon i\pi r} z^r \quad \text{if} \quad z \in H_\epsilon \tag{8.3}$$

where $\epsilon = \pm$ and H_ϵ is the upper (resp., lower) half plane. (We shall typically use this for $z = z_1 - z_2$.) While $(z_1 z_2)^r$ and $z_1^r z_2^r$ are not always equal, they are in fact equal if

$$\operatorname{Re} z_1 > 0, \quad \operatorname{Re} z_2 > 0. \tag{8.4}$$

Thus for $|z_1| > |z_2| > 0$ and $\operatorname{Re} z_1 > 0$ (in which case $z_1 - z_2 \not\leq 0$),

$$(z_1 - z_2)^r = z_1^r (1 - z_2/z_1)^r. \tag{8.5}$$

We shall need the following observation relating a formal binomial series and a holomorphic function on a certain domain:

Remark 8.1: The formal series

$$(z_1 - z_2)^r = z_1^r (1 - z_2/z_1)^r = \sum_{m \geq 0} \binom{r}{m} (-1)^m z_1^{r-m} z_2^m \tag{8.6}$$

converges to the holomorphic function $(z_1 - z_2)^r$ in the domain

$$\{(z_1, z_2) \in (\mathbf{C}^\times)^2 \,|\, |z_1| > |z_2| > 0, \ \operatorname{Re} z_1 > 0\}. \tag{8.7}$$

For a permutation $\sigma = (i_1 \cdots i_n)$ of $(1 \cdots n)$, set

$$R_\sigma = \{(z_1, ..., z_n) \in (\mathbf{C}^\times)^n \,|\, |z_{i_1}| > \cdots > |z_{i_n}| > 0, \ \operatorname{Re} z_l > 0\} \tag{8.8}$$

$$U_\sigma = \{(z_1, ..., z_n) \in (\mathbf{C}^\times)^n | z_l \not\leq 0, \ z_{i_j} \not\leq z_{i_k} \ \text{if} \ j < k\} \supset R_\sigma. \tag{8.9}$$

For $i = 1, ..., n-1$, also set

$$\begin{aligned} U(i, \epsilon) = \{(z_1, ..., z_n) \in (\mathbf{C}^\times)^n | z_l \not\leq 0, \ z_j \not\leq z_k \text{ if } j < k, \\ z_i - z_{i+1} \in H_\epsilon\} \subset U_1 \cap U_{(i,i+1)}, \end{aligned} \tag{8.10}$$

where $\sigma = 1$ is the trivial permutation and $\sigma = (i, i+1)$ is the transposition. The domains $U(i,+)$ and $U(i,-)$ are clearly disjoint.

The following result is an immediate consequence of Proposition 7.12, in the setting of either an algebra or a module (recall Remark 7.19):

Proposition 8.2 *(a) In the notation of Proposition 7.12, the formal series*

$$\langle v', Y(v_{i_1}, z_{i_1}) \cdots Y(v_{i_n}, z_{i_n}) v \rangle \tag{8.11}$$

converges to the (single-valued) holomorphic function

$$f(z_1, ..., z_n) \prod_{j<k} (z_{i_j} - z_{i_k})^{-(g_{i_j}, g_{i_k})} \prod_{l=1}^{n} z_l^{-(g_l, h)} \tag{8.12}$$

on the domain R_σ. This function has a single-valued analytic continuation on the domain U_σ.
(b) As holomorphic functions on the domain $U(i, \epsilon)$ ($i = 1, ..., n-1, \epsilon = \pm$),

$$\begin{aligned} &\langle v', Y(v_1, z_1) \cdots Y(v_{i-1}, z_{i-1}) Y(v_{i+1}, z_{i+1}) \cdot \\ &\quad \cdot Y(v_i, z_i) Y(v_{i+2}, z_{i+2}) \cdots Y(v_n, z_n) v \rangle \\ &= (e^{2\pi i/2T})^{(g_i, g_{i+1})\epsilon T} \langle v', Y(v_1, z_1) \cdots Y(v_n, z_n) v \rangle. \quad \square \end{aligned} \tag{8.13}$$

In order to interpret this result in terms of representations of the braid group, we adapt the framework of [TK], Section 5.1. We consider the complex manifold

$$X_n = \{(z_1, ..., z_n) \in \mathbf{C}^n | z_i \neq z_j \text{ if } i \neq j\}. \tag{8.14}$$

Then the n-th symmetric group S_n acts on the right on the manifold X_n by: $(z_1, ..., z_n)\sigma = (z_{\sigma(1)}, ..., z_{\sigma(n)})$ for $\sigma \in S_n$. The braid group B_n is the fundamental group $\pi_1(X_n/S_n)$ of the homogeneous space X_n/S_n. Introduce the subset I_1 of X_n defined by

$$\begin{aligned} I_1 &= \{(x_1, ..., x_n) \in \mathbf{R}^n | x_1 > \cdots > x_n > 0\} \\ &= R_1 \cap \mathbf{R}^n \subset X_n \end{aligned} \tag{8.15}$$

(recall (8.8)). It is well known that B_n has generators $b_1, ..., b_{n-1}$ with defining relations

$$\begin{aligned} b_i b_j &= b_j b_i \ (|i-j| > 1) \\ b_i b_{i+1} b_i &= b_{i+1} b_i b_{i+1} \ (i = 1, ..., n-2), \end{aligned} \tag{8.16}$$

where b_i represented by the path

$$\zeta_i(t) = (\zeta_i^1(t), ..., \zeta_i^n(t)) \in X_n \quad (t \in [0,1]) \tag{8.17}$$

(which defines a loop in X_n/S_n) given by

$$\begin{aligned} \zeta_i^k(t) &= z_k \ (k \neq i, i+1) \\ \zeta_i^i(t) &= \frac{z_i + z_{i+1}}{2} + e^{i\pi t}\frac{z_i - z_{i+1}}{2} \\ \zeta_i^{i+1}(t) &= \frac{z_i + z_{i+1}}{2} - e^{i\pi t}\frac{z_i - z_{i+1}}{2} \end{aligned} \tag{8.18}$$

for $(z_1, ..., z_n) \in I_1$. Let $\tilde{X}_n$ denote the universal covering space of X_n (or of X_n/S_n). A complex-valued holomorphic function φ on $\tilde{X}_n$ (or equivalently, a multiple-valued holomorphic function on X_n or on X_n/S_n) is uniquely determined by its *principal branch* – the restriction of φ to a given subset of $\tilde{X}_n$ homeomorphic to I_1 under the covering map. The group B_n acts naturally on the space of such functions φ, in such a way that the principal branch of $b_i \cdot \varphi$ is obtained by analytic continuation of the principal branch of φ along the path ζ_i.

It will be convenient now to restrict our attention to an algebra as opposed a module. Let $g_1, ..., g_n \in G$, let $v_j \in V^{g_j}$ and set

$$\mathbf{v} = (v', v_1, ..., v_n). \tag{8.19}$$

Recalling Remark 7.14, we now regard

$$Y_{\mathbf{v}}(z_1, ..., z_n) = \langle v', Y(v_1, z_1) \cdots Y(v_n, z_n)\mathbf{1}\rangle \tag{8.20}$$

as the multi-valued (algebraic) holomorphic function on X_n obtained by analytically continuing (8.12) (for $\sigma = 1$ and $v = \mathbf{1}$) from the domain R_1. Its principal branch is given by the restriction of (8.12) to I_1. Then by Proposition 8.2 (b), with $\epsilon = +$ and $(z_1, ..., z_n)$ taken to be $\zeta_j(t)$ $(0 < t < 1)$, with $t \to 1$, the principal branch of $b_j \cdot Y_{\mathbf{v}}$ is given by:

$$(b_j \cdot Y_{\mathbf{v}})(z_1, ..., z_n) = e^{-i\pi(g_j, g_{j+1})} Y_{(j,j+1)\cdot \mathbf{v}}(z_1, ..., z_n) \tag{8.21}$$

for $j = 1, ..., n-1$ and $(z_1, ..., z_n) \in I_1$, where the transposition $(j, j+1)$ acts naturally on $\mathbf{v}$. Thus

$$b_j \cdot Y_{\mathbf{v}} = e^{-i\pi(g_j, g_{j+1})} Y_{(j,j+1)\cdot \mathbf{v}}. \tag{8.22}$$

Let M be an S_n-module, possibly infinite-dimensional. Then B_n acts on M via the natural surjection $B_n \to S_n$, and B_n acts on the space of all M-valued functions ψ on $\tilde{X}_n$ (ψ not assumed holomorphic in any sense) by:

$$(\tau \cdot \psi)(x) = \tau \cdot \psi(x \cdot \tau) \tag{8.23}$$

for $\tau \in B_n$, $x \in \tilde{X}_n$, where $x \cdot \tau$ denotes the natural right covering-transformation action of B_n on each fiber of the covering $\tilde{X}_n \to X_n/S_n$.

Now we assume that

$$g_1 = \cdots = g_n = g \tag{8.24}$$

and we set

$$V(g) = (V' \otimes (\otimes^n V^g))^*, \tag{8.25}$$

an S_n-module in a natural way: For $\sigma \in S_n, f \in V(g), v' \in V'$ and $v_j \in V^g$,

$$(\sigma \cdot f)(v' \otimes v_1 \otimes \cdots \otimes v_n) = f(v' \otimes v_{\sigma(1)} \otimes \cdots \otimes v_{\sigma(n)}). \tag{8.26}$$

Consider the $V(g)$-valued function Y on $\tilde{X}_n$ (i.e., the multi-valued function on X_n) given by:

$$Y(z_1, ..., z_n)(v' \otimes v_1 \otimes \cdots \otimes v_n) = Y_{\mathbf{v}}(z_1, ..., z_n) \tag{8.27}$$

(recall (8.20)), which is multilinear in $\mathbf{v}$. Then by (8.22) (with b_j^{-1} in place of b_j),

$$b_j \cdot Y = e^{-i\pi(g,g)}Y, \tag{8.28}$$

under the natural action (8.23) of B_n. Consider the one-dimensional function space

$$W(g) = \mathbf{C}Y. \tag{8.29}$$

Then we have:

Proposition 8.3 *The space $W(g)$ is a natural B_n-module, and for $j = 1, ..., n-1$, b_j acts on $W(g)$ as multiplication by $e^{-i\pi(g,g)}$, where (g, g) is an element of $(\frac{1}{T}\mathbf{Z})/2\mathbf{Z}$.* □

Remark 8.4: This representation also gives a representation of the Hecke algebra $H_n(q)$ for $q = e^{-i\pi(g,g)}$, which is defined as the associative algebra with generators $T_1, ..., T_{n-1}$ subject to the defining relations

$$T_iT_{i+1}T_i = T_{i+1}T_iT_{i+1} \quad \text{for} \quad i = 1, ..., n-2 \tag{8.30}$$

$$T_iT_j = T_jT_i \quad \text{for} \quad |i-j| > 1 \tag{8.31}$$

$$(T_i - q)(T_i + 1) = 0. \tag{8.32}$$

Chapter 9

Generalized vertex algebras and duality

Now we return to the setting of Chapter 5. The axioms for a generalized vertex operator algebra do not quite apply to the structure of Ω_*, the vacuum space (3.29) (for the Heisenberg algebra $(\hat{\mathbf{h}}_*)_{\mathbf{Z}}$) equipped with the relative vertex operators Y_*. In fact, the finite-dimensionality axiom (6.4) and the boundedness axiom (6.5) fail to hold in general, since L is not necessarily positive definite. Moreover, the Jacobi identity (5.11) (Theorem 5.1) differs from the Jacobi identity (6.12) by the insertion of the factor $c(\bar{a}, \bar{b})$. Of course, a suitable group G would also have to be defined. We would like to extend our axioms so as to include this structure. Then the corresponding generalization of the duality arguments of Chapter 7 would provide an alternate duality-based proof of the Jacobi identity (5.11) (recall Remark 7.18), and in particular, of Theorems 8.6.1 and 8.8.23 of [FLM3]. (The arguments in Chapter 7 above, in the Appendix of [FLM3] and in [FHL] are not sufficiently general to give a duality-based proof of Theorem 8.8.23 of [FLM3] – the special case of (5.11) in which $\mathbf{h}_* = 0$ and $\langle \bar{a}, \bar{b} \rangle \in \mathbf{Z}$ – because the rational lattice in that theorem is not assumed even or integral, and has an alternating form $c(\cdot, \cdot)$; in the proof of Proposition 7.16, the vector v_3 is arbitrary.)

In this chapter we shall consider a more general algebraic structure, "generalized vertex algebras," which include the setting of Theorem 5.1 and the generalized vertex operator algebras defined in Chapter 6, together with a variant of what Borcherds calls "vertex algebras" [Bor], and for which the duality and related results in Chapter 7 remain valid with expected modifications. As an application, one can give a new proof of Theorem 5.1 of this monograph, and therefore a new proof of Theorems 8.6.1 and 8.8.23 of [FLM3]. We remark that even in the standard case in which L is even and the commutator map is the usual one, the structure of the generalized vertex algebra V_L can incorporate more information than the vertex algebra structure, namely, the information of the central extension of L.

After defining the notion of generalized vertex algebra, we specify the exact sense in which the notion of generalized vertex operator algebra (Chapter 6) is a special case, and we also identify another special case – that of "vertex algebra," a variant of Borcherds' notion [Bor]. We also discuss some basic properties of generalized vertex algebras and some related notions, especially the notion of module and useful variants of this notion. Then we show that under appropriate hypotheses the vacuum space Ω_* studied in Chapter 5 has the structure of a generalized vertex algebra, when a suitable group G is defined. The special situation in which $\mathbf{h}_* = 0$ (so that $\Omega_* = V_L$) is significant; it includes the setting of Theorem 8.8.23 of [FLM3]. We mention various families of interesting special cases, including an important setting involving an even sublattice L_0 of our lattice L, with certain hypotheses, giving rise to a vertex algebra acting on a family of irreducible modules (see Remark 9.10); this setting will be used in Chapter 12 as the starting point for constructing a basic example of an abelian intertwining algebra. The results on duality (Chapter 7) and monodromy (Chapter 8) are carried over from the earlier setting of generalized vertex operator algebras. The proofs being virtually the same as before, we state only the results. Only one subtle change is necessary: An additional hypothesis (see (9.39)), which holds in the case of the basic example built from Chapter 5, is needed for the proof of generalized rationality and commutativity for products of more than two vertex operators. Using the result that the Jacobi identity follows from generalized commutativity, we sketch an alternative proof of the Jacobi identity (5.11), and hence of the special cases of this result, Theorems 8.6.1 and 8.8.23 of [FLM3]. We observe that essentially the entire discussion of duality applies, with obvious modifications, to modules. Finally, we discuss the special case of "vertex superalgebras."

First we give our definition of generalized vertex algebra. Roughly speaking, the axioms for a generalized vertex algebra are the same as those for a generalized vertex operator algebra (Chapter 6) except that the grading-restriction axioms (6.4) and (6.5) are removed and the Jacobi identity (6.12) is modified by the insertion of the factor $c(g, h)$.

Let T be a positive integer. Let G be an abelian group (not necessarily finite) and suppose that $(\cdot,\cdot)$ is a symmetric $(\frac{1}{T}\mathbf{Z})/2\mathbf{Z}$-valued $\mathbf{Z}$-bilinear form (not necessarily nondegenerate) on G :

$$(9.1) \qquad (g,h) \in \left(\frac{1}{T}\mathbf{Z}\right)/2\mathbf{Z} \quad \text{for} \quad g,h \in G$$

and that $c(\cdot,\cdot)$ is an alternating $\mathbf{Z}$-bilinear form on G with values in the multiplicative group $\mathbf{C}^\times$:

$$(9.2) \qquad c(g,h) \in \mathbf{C}^\times$$

$$(9.3) \qquad c(g_1+g_3,g_2) = c(g_1,g_2)c(g_3,g_2)$$

$$(9.4) \qquad c(g_1,g_2+g_3) = c(g_1,g_2)c(g_1,g_3)$$

$$(9.5) \qquad c(g,g) = 1$$

for $g,h,g_i \in G$. *A generalized vertex algebra* of *level* T associated with the group G and the forms $(\cdot,\cdot)$ and $c(\cdot,\cdot)$ is a vector space with two gradations:

$$(9.6) \qquad V = \coprod_{n\in\frac{1}{T}\mathbf{Z}} V_n = \coprod_{g\in G} V^g; \quad \text{for} \quad v \in V_n, \quad n = \operatorname{wt} v;$$

such that

$$(9.7) \qquad V^g = \coprod_{n\in\frac{1}{T}\mathbf{Z}} V_n^g \quad (\text{where } V_n^g = V_n \bigcap V^g) \quad \text{for} \quad g \in G,$$

equipped with a linear map

$$(9.8) \qquad \begin{array}{rcl} V & \to & (\operatorname{End} V)[[z^{\frac{1}{T}}, z^{-\frac{1}{T}}]] \\ v & \mapsto & Y(v,z) = \sum_{n\in\frac{1}{T}\mathbf{Z}} v_n z^{-n-1} \quad (v_n \in \operatorname{End} V) \end{array}$$

and with two distinguished vectors $\mathbf{1} \in V_0^0$, $\omega \in V_2^0$, satisfying the following conditions for $g,h \in G$, $u,v \in V$ and $l \in \frac{1}{T}\mathbf{Z}$:

$$(9.9) \qquad u_l V^h \subset V^{g+h} \quad \text{if} \quad u \in V^g;$$

$$(9.10) \qquad u_l v = 0 \quad \text{if } l \text{ is sufficiently large;}$$

$$(9.11) \qquad Y(\mathbf{1},z) = 1;$$

$$(9.12) \qquad Y(v,z)\mathbf{1} \in V[[z]] \quad \text{and} \quad \lim_{z\to 0} Y(v,z)\mathbf{1} = v;$$

$$(9.13) \qquad Y(v,z)|_{V^h} = \sum_{n\equiv(g,h) \bmod \mathbf{Z}} v_n z^{-n-1} \quad \text{if} \quad v \in V^g$$

(i.e., $n + 2\mathbf{Z} \equiv (g, h) \bmod \mathbf{Z}/2\mathbf{Z}$);

$$\begin{aligned} & z_0^{-1}\left(\frac{z_1 - z_2}{z_0}\right)^{(g,h)} \delta\left(\frac{z_1 - z_2}{z_0}\right) Y(u, z_1)Y(v, z_2) \\ (9.14) \quad & - c(g, h) z_0^{-1}\left(\frac{z_2 - z_1}{z_0}\right)^{(g,h)} \delta\left(\frac{z_2 - z_1}{-z_0}\right) Y(v, z_2)Y(u, z_1) \\ & = z_2^{-1}\delta\left(\frac{z_1 - z_0}{z_2}\right) Y(Y(u, z_0)v, z_2)\left(\frac{z_1 - z_0}{z_2}\right)^{-g} \end{aligned}$$

(the generalized Jacobi identity) if $u \in V^g$, $v \in V^h$, where

$$(9.15) \qquad \left(\frac{z_1 - z_0}{z_2}\right)^{-g} \delta\left(\frac{z_1 - z_0}{z_2}\right) \cdot w = \left(\frac{z_1 - z_0}{z_2}\right)^{-(g,k)} \delta\left(\frac{z_1 - z_0}{z_2}\right) w$$

for $w \in V^k$, $k \in G$ (note that all three terms in (9.14) are well defined);

$$(9.16) \quad [L(m), L(n)] = (m-n)L(m+n) + \frac{1}{12}(m^3 - m)\delta_{m+n,0}(\operatorname{rank} V)$$

for $m, n \in \mathbf{Z}$, where

$$(9.17) \quad L(n) = \omega_{n+1} \text{ for } n \in \mathbf{Z}, \text{ i.e., } Y(\omega, z) = \sum_{n \in \mathbf{Z}} L(n) z^{-n-2}$$

and

$$(9.18) \qquad \operatorname{rank} V \in \mathbf{Q};$$

$$(9.19) \qquad L(0)v = nv = (\operatorname{wt} v)v \quad \text{for} \quad n \in \frac{1}{T}\mathbf{Z}, \quad v \in V_n;$$

$$(9.20) \qquad \frac{d}{dz} Y(v, z) = Y(L(-1)v, z).$$

We denote the generalized vertex algebra just defined by

$$(9.21) \qquad (V, Y, \mathbf{1}, \omega, T, G, (\cdot, \cdot), c(\cdot, \cdot))$$

or briefly, by V. As usual, the expressions $Y(v, z)$ are called *(generalized) vertex operators*. Note that we can have $\omega = 0$, and $V = 0$ or $V = \mathbf{C1}$. The considerations of Remark 6.1 still hold for the form $(\cdot, \cdot)$ on G.

Remark 9.1: If $(g, h) \in \mathbf{Z}/2\mathbf{Z}$ in (9.14), then the generalized Jacobi identity (9.14) reduces to:

$$z_0^{-1}\delta\left(\frac{z_1 - z_2}{z_0}\right) Y(u, z_1)Y(v, z_2)$$

$$-c(g,h)(-1)^{(g,h)}z_0^{-1}\delta\left(\frac{z_2-z_1}{-z_0}\right)Y(v,z_2)Y(u,z_1)$$

$$= z_2^{-1}\delta\left(\frac{z_1-z_0}{z_2}\right)Y(Y(u,z_0)v,z_2)\left(\frac{z_1-z_0}{z_2}\right)^{-g}. \tag{9.22}$$

Furthermore, if $c(g,h) = (-1)^{(g,h)}$ and $(g,G) \subset \mathbf{Z}/2\mathbf{Z}$, we get the ordinary Jacobi identity (5.28).

Remark 9.2: The case in which the abelian group G is finite, the form $(\cdot,\cdot)$ is nondegenerate on G and $c(\cdot,\cdot) = 1$ together with the axioms (6.4) and (6.5) gives precisely the definition of the generalized vertex operator algebra $(V, Y, \mathbf{1}, \omega, T, G, (\cdot,\cdot))$ (recall (6.19)).

Remark 9.3: The consequences (6.20)-(6.28) and Remarks 6.2 and 6.3 remain valid for a generalized vertex algebra.

Remark 9.4: The case $T = 1$, $G = 0$ in the definition of generalized vertex algebra gives the definition of *vertex algebra* (a variant of Borcherds' notion [Bor]).

Remark 9.5: The analogue of Remark 6.5 holds here: If V^0 satisfies the condition (6.30), then $(V^0, Y, \mathbf{1}, \omega)$ is a vertex algebra, in the sense of Remark 9.4. Also, the axiom (9.13) follows from the assertion (6.31) together with the other axioms for a generalized vertex algebra. Formulas (6.30) and (6.31) hold for the generalized vertex algebra of Theorem 9.8 below.

Remark 9.6: An *automorphism* of the generalized vertex algebra (9.21) is a pair (ρ, σ) as in the case of the generalized vertex operator algebra (6.19) (recall (6.32)-(6.35)), except that we add the natural condition that σ should preserve $c(\cdot,\cdot)$. As in the case of a generalized vertex operator algebra, G acts naturally as a group of automorphisms of V, and for any group K of automorphisms of V, the group $K \ltimes G$ acts as a group of automorphisms of V.

Remark 9.7: We have the expected notion of module: Let $V = (V, Y, \mathbf{1}, \omega, T, G, (\cdot,\cdot), c(\cdot,\cdot))$ be a generalized vertex algebra. In the definition of module for a generalized vertex operator algebra, if we remove (6.46)-(6.47) and insert the factor $c(g,h)$ in the Jacobi identity (6.53), we have a module (W, Y) for the generalized vertex algebra V. As before, we have the notion of *adjoint module*, and formulas (6.59)-(6.63) and Remark 6.11 hold in the present context. More generally, as in Remark 6.15, given a G-set S and a $(\frac{1}{T}\mathbf{Z})/2\mathbf{Z}$-valued function $(\cdot,\cdot)$ on $G \times S$ satisfying (6.64), then we have the corresponding notion of ($\mathbf{Q}$-graded) V-module $W = (W, Y, S, (\cdot,\cdot))$, with the expected properties. (One could also generalize the $\mathbf{Q}$-grading to a $\mathbf{C}$-grading, as in Remark 6.16.)

Next we show that the vacuum space Ω_* (see Chapters 3 and 5) has the structure of a generalized vertex algebra. Here we work under Assumptions 2.1-2.4 and the rational compatibility assumption (5.30). The case $\mathbf{h}_* = 0$ (which gives $\Omega_* = V_L$) includes the setting of Theorem 8.8.23 of [FLM3].

Let T be the smallest positive integer such that

$$\langle \alpha', \alpha' \rangle \in \frac{2}{T}\mathbf{Z} \quad \text{for} \quad \alpha \in L \tag{9.23}$$

(cf. (5.34)-(5.35)). Recall the ν-invariant alternating $\mathbf{Z}$-bilinear form $c(\cdot,\cdot)$ on L ((3.26)-(3.28)). Let R be the radical of c:

$$R = \{\alpha \in L | c(\alpha, \beta) = 1 \text{ for } \beta \in L\}. \tag{9.24}$$

Recalling (5.33) and (5.40) we set

$$H = R \cap 2(L')^{\circ} \tag{9.25}$$

(which is different from the group H in (5.44)) and we define an abelian group

$$G = L/H \tag{9.26}$$

(this group is also different from that defined in (5.43)). Then G is finite and ν acts naturally as an automorphism of G. The subgroup H is chosen so that the bilinear forms (9.27) and (9.28) on G below are well defined. As in (5.45) we define a $(\frac{1}{T}\mathbf{Z})/2\mathbf{Z}$-valued $\mathbf{Z}$-bilinear form $(\cdot,\cdot)$ on G by:

$$\begin{aligned} G \times G &\to \left(\tfrac{1}{T}\mathbf{Z}\right)/2\mathbf{Z} \\ (\alpha + H, \beta + H) &\mapsto (\alpha + H, \beta + H) = -\langle \alpha', \beta' \rangle + 2\mathbf{Z} \end{aligned} \tag{9.27}$$

for $\alpha, \beta \in L$. This form is symmetric and ν-invariant, but not necessarily nondegenerate. We also regard $c(\cdot,\cdot)$ as a ν-invariant alternating

Z-bilinear form on G:

$$\begin{aligned} G \times G &\rightarrow \langle \omega_q \rangle \subset \mathbf{C}^\times \\ (\alpha + H, \beta + H) &\mapsto c(\alpha + H, \beta + H) = c(\alpha, \beta) \end{aligned} \tag{9.28}$$

for $\alpha, \beta \in L$ (recall (3.26)). For $g \in G$ set

$$\Omega_*^g = \sum_{b \in \hat{L},\ \bar{b}+H=g} S((\hat{\mathbf{h}}_*^\perp)^-) \otimes \iota(b). \tag{9.29}$$

(recall (3.33)). Then we have the decomposition

$$\Omega_* = \coprod_{g \in G} \Omega_*^g, \tag{9.30}$$

and

$$Y_*(u, z)\Omega_*^g \subset \Omega_*^{g+h}[[z^{\frac{1}{T}}, z^{-\frac{1}{T}}]] \tag{9.31}$$

for $g, h \in G$ and $u \in \Omega_*^h$.

Theorem 9.8 *Let L be a rational lattice satisfying Assumptions 2.1-2.4 and the rational compatibility condition (5.30). Then*

$$(\Omega_*, Y_*, \mathbf{1}, \omega, T, G, (\cdot,\cdot), c(\cdot,\cdot)) \tag{9.32}$$

is a generalized vertex algebra with G finite and with rank equal to $\dim \mathbf{h}_*^\perp$, and $\hat{\nu}$ acts naturally as an automorphism. In particular, the group $\langle \hat{\nu} \rangle \ltimes G$ acts as a group of automorphisms of the generalized vertex algebra Ω_*. □

Remark 9.9: If L is even, $c(\alpha, \beta) = (-1)^{\langle \alpha, \beta \rangle}$ for $\alpha, \beta \in L$ and $\mathbf{h}_* = 0$, then $\Omega_* = V_L$, $Y_* = Y$ (as in [FLM3] and in Remark 3.2 and Corollary 5.4 above), $T = 1$, $G = L/H$ ($H = L \cap 2L^\circ$) and $c(g, h) = (-1)^{(g,h)}$ for $g, h \in G$. By Remarks 9.1 and 9.4, V_L has the structure of a vertex algebra, but we emphasize that the generalized vertex algebra structure

$$(V_L, Y, \mathbf{1}, \omega, 1, G, (\cdot,\cdot), c(\cdot,\cdot)) \tag{9.33}$$

includes the information of the central extension of L (cf. Remarks 6.4 and 6.9). Also note the case $\mathbf{h}_* = \mathbf{h}$.

Remark 9.10: Extending the setting of Remark 9.9, let L again be general, and let L_0 be an even sublattice of L such that $L \subset L_0^\circ$ and $c(\alpha,\beta) = (-1)^{\langle\alpha,\beta\rangle}$ for $\alpha,\beta \in L_0$, with q even, as in Corollary 5.5. Assume also that $\mathbf{h}_* = 0$, so that $Y_* = Y$ (recall Remark 3.2). Also recall the notation (5.53)-(5.58). By Theorem 5.1 and Propositions 5.6, 5.7 and 5.10, $(V(0), Y, \mathbf{1}, \omega)$ is a vertex algebra (in the sense of Remark 9.4) and $(V(i), Y)$ is an irreducible $V(0)$-module for $i \in L/L_0$.

Remark 9.11: Theorem 5.1 with $\mathbf{h}_* = 0$ suggests another example of generalized vertex algebra, this time with G infinite: Take T as in (9.23), $G = L$, $(\alpha,\beta) = -\langle\alpha,\beta\rangle + 2\mathbf{Z}$ for $\alpha,\beta \in L$, $c(\cdot,\cdot)$ as in Theorem 5.1, $V = V_L$, and $Y, \mathbf{1}$ and ω as in the earlier chapters. Note that formula (6.31) holds in this case. Also note that the grading of each G-homogeneous subspace of V is bounded below.

We now discuss duality for a generalized vertex algebra. Suitable variants of all the arguments and results in Chapters 7 and 8 remain valid for a generalized vertex algebra $(V, Y, \mathbf{1}, \omega, T, G, (\cdot,\cdot), c(\cdot,\cdot))$, with the proper insertion of the factor $c(\cdot,\cdot)$, and with an additional hypothesis stated below. We proceed to formulate the results. Adopting the notational conventions of Remarks 7.1 and 7.2, let $g, h \in G$ and $u \in V^g, v \in V^h$. We have the new *generalized commutator relation* for the vertex operators $Y(u, z_1)$ and $Y(v, z_2)$:

$$\begin{aligned}
&(z_1 - z_2)^{(g,h)} Y(u, z_1) Y(v, z_2) z_1^g z_2^h \\
&\quad - c(g,h)(z_2 - z_1)^{(g,h)} Y(v, z_2) Y(u, z_1) z_1^g z_2^h \\
&= \sum_{n=0}^{N} (-1)^n Y\Big(\big(Y(u, z_1 - z_2)(z_1 - z_2)^{(g,h)+n} \\
&\quad - Y(u, -z_2 + z_1)(-z_2 + z_1)^{(g,h)+n}\big) v, z_2\Big) z_1^{-n} z_2^{g+h} \binom{-g}{n},
\end{aligned} \tag{9.34}$$

where N is a large enough nonnegative integer. If $c(\cdot,\cdot) = 1$, this is the generalized commutator relation (7.5). Now generalized commutativity becomes accordingly:

Proposition 9.12 *(a)* **(generalized rationality of products)** *For $g_1, g_2, h \in G$ and $v_1 \in V^{g_1}, v_2 \in V^{g_2}, v \in V^h, v' \in V'$, the formal series*

$$\langle v', Y(v_1, z_1) Y(v_2, z_2) v\rangle (z_1 - z_2)^{(g_1,g_2)} z_1^{(g_1,h)} z_2^{(g_2,h)},$$

which is well defined and which involves only integral powers of z_1 *and* z_2*, lies in the image of the map* ι_{12} :

$$\begin{aligned}(9.35)\qquad &\langle v', Y(v_1, z_1)Y(v_2, z_2)v\rangle (z_1 - z_2)^{(g_1,g_2)} z_1^{(g_1,h)} z_2^{(g_2,h)} \\ &= \iota_{12} f(z_1, z_2),\end{aligned}$$

where the (uniquely determined) element $f \in \mathbf{C}[z_1, z_2]_S$ *is of the form*

$$(9.36)\qquad f(z_1, z_2) = g(z_1, z_2)/z_1^{r_1} z_2^{r_2} (z_1 - z_2)^t$$

for some $g \in \mathbf{C}[z_1, z_2]$ *and* $r_1, r_2, t \in \mathbf{Z}$*. The integer* t *depends only on* v_1 *and* v_2*; it is independent of* v *and* v'*.*
(b) **(generalized commutativity)** *We also have*

$$\begin{aligned}(9.37)\qquad &c(g_1, g_2)\langle v', Y(v_2, z_2)Y(v_1, z_1)v\rangle (z_2 - z_1)^{(g_1,g_2)} z_1^{(g_1,h)} z_2^{(g_2,h)} \\ &= \iota_{21} f(z_1, z_2)\end{aligned}$$

(in particular, the left-hand side is well defined), that is,

$$\begin{aligned}(9.38)\qquad &\text{“}Y(v_1, z_1)Y(v_2, z_2)(z_1 - z_2)^{(g_1,g_2)} z_1^{g_1} z_2^{g_2} \quad \text{agrees with} \\ &c(g_1, g_2)Y(v_2, z_2)Y(v_1, z_1)(z_2 - z_1)^{(g_1,g_2)} z_1^{g_1} z_2^{g_2}\end{aligned}$$

as operator-valued rational functions.” □

If $c(\cdot, \cdot) = 1$, this is exactly the generalized rationality and commutativity for a generalized vertex operator algebra presented in Chapter 7. The obvious analogue of Remark 7.7 holds here. The generalized associator relation (7.11) holds with the insertion of the factor $c(g, h)$ on the right-hand side, and thus generalized associativity and the related results remain valid without change for generalized vertex algebras. More precisely, we have:

Proposition 9.13 *The assertions of Propositions 7.8, 7.9, 7.11 and Corollary 7.10 hold without change for a generalized vertex algebra.* □

Recall from Proposition 7.12 that generalized rationality and commutativity hold for an arbitrary number of vertex operators for a generalized vertex operator algebra, but the argument does not hold for a generalized vertex algebra unless we make an additional assumption: For $v_1, ..., v_n, v \in V$ $(n > 0)$, assume that there exists $r \in \frac{1}{T}\mathbf{Z}$ such that

$$\text{(9.39)}\quad \begin{array}{c}\text{the coefficient of each monomial in } z_1, ..., z_n \\ \text{in } Y(v_1, z_1) \cdots Y(v_n, z_n)v \text{ lies in } \coprod_{m\in\frac{1}{T}\mathbf{Z},\ m>r} V_m.\end{array}$$

Remark 9.14: The hypothesis (9.39) holds for an arbitrary generalized vertex algebra for $n = 1$, by (9.10) and Remarks 9.3 and 6.3.

Remark 9.15: If the boundedness axiom (6.5) holds, then (9.39) obviously holds.

Remark 9.16: For the generalized vertex algebra Ω_* of Theorem 9.8 or the algebra V_L of Remark 9.11, the boundedness axiom (6.5) fails to hold in general, but the hypothesis (9.39) is clearly true.

Now we repeat the proof of Proposition 7.12, using the hypothesis (9.39) in place of the boundedness condition (6.5) in studying (7.44), and keeping in mind the alternating property of $c(\cdot,\cdot)$ (recall (9.2)-(9.5)). The result is the following generalized rationality and commutativity for several vertex operators:

Proposition 9.17 (generalized rationality and commutativity) *Assume (9.39). For $g_1, ..., g_n, h \in G$ and $v_1 \in V^{g_1}, ..., v_n \in V^{g_n}, v \in V^h, v' \in V'$ and any permutation $(i_1 \cdots i_n)$ of $(1 \cdots n)$, the formal series*

$$\left(\prod_{j<k,\ i_j>i_k} c(g_{i_k}, g_{i_j})\right) \langle v', Y(v_{i_1}, z_{i_1}) \cdots Y(v_{i_n}, z_{i_n})v\rangle \cdot$$

$$\cdot \prod_{j<k} (z_{i_j} - z_{i_k})^{(g_{i_j}, g_{i_k})} \prod_{l=1}^{n} z_l^{(g_l, h)},$$

which is well defined and which involves only integral powers of each z_i, lies in the image of the map $\iota_{i_1 \cdots i_n}$:

$$\text{(9.40)}\quad \left(\prod_{j<k,\ i_j>i_k} c(g_{i_k}, g_{i_j})\right) \langle v', Y(v_{i_1}, z_{i_1}) \cdots Y(v_{i_n}, z_{i_n})v\rangle \cdot \\ \cdot \textstyle\prod_{j<k} (z_{i_j} - z_{i_k})^{(g_{i_j}, g_{i_k})} \prod_{l=1}^{n} z_l^{(g_l, h)} = \iota_{i_1 \cdots i_n} f(z_1, ..., z_n),$$

where the (uniquely determined) element $f \in \mathbf{C}[z_1, ..., z_n]_S$ is independent of the permutation and is of the form

$$(9.41) \qquad f(z_1, ..., z_n) = g(z_1, ..., z_n) / \prod_{i=1}^{n} z_i^{r_i} \prod_{j<k} (z_j - z_k)^{t_{jk}}$$

for some $g \in \mathbf{C}[z_1, ..., z_n]$ and $r_i, t_{jk} \in \mathbf{Z}$. The integer t_{jk} depends only on v_j and v_k; it is independent of v_l ($l \neq j, k$), v and v'. In particular,

$$\text{“}Y(v_1, z_1) \cdots Y(v_n, z_n) \prod_{j<k} (z_j - z_k)^{(g_j, g_k)} \prod_{l=1}^{n} z_l^{g_l} \text{ agrees with}$$

$$(9.42) \qquad \left(\prod_{j<k,\ i_j > i_k} c(g_{i_k}, g_{i_j}) \right) Y(v_{i_1}, z_{i_1}) \cdots Y(v_{i_n}, z_{i_n}) \cdot$$
$$\cdot \textstyle\prod_{j<k} (z_{i_j} - z_{i_k})^{(g_{i_j}, g_{i_k})} \prod_{l=1}^{n} z_l^{g_l}$$

as operator-valued rational functions.” □

The remaining arguments in Chapter 7 and those in Chapter 8 carry over. The expression (8.11) is multiplied by the extra factor in (9.40), and the left-hand side of (8.13) and the right-hand sides of (8.21) and (8.22) are multiplied by $c(g_i, g_{i+1})$. But because of the specialization (8.24), the final conclusions on braid group representations in Chapter 8 remain the same, and we have:

Proposition 9.18 *Under the hypothesis (9.39), the assertions of Propositions 7.16 and 8.3, Corollary 7.15 and Remarks 7.13, 7.14, 7.17 and 8.4 are valid without change for a generalized vertex algebra.* □

Remark 9.19: By the assertion of Remark 7.17, the generalized Jacobi identity (9.14) is equivalent to generalized commutativity (Proposition 9.17) for three vertex operators applied to v=**1** and the $L(-1)$-bracket formula (6.20). This suggests a new proof of Theorem 5.1 (strictly speaking, in the slightly restricted setting of Theorem 9.8), and hence a new proof of Theorem 8.8.23 of [FLM3], and in particular, of Theorem 8.6.1 of [FLM3]. Briefly, by the argument in the proof of Proposition A.2.1 of [FLM3] we get generalized rationality and commutativity for any number of vertex operators, and the $L(-1)$-bracket formula follows from the easy results

$$(9.43) \qquad [L(-1), \alpha(z)] = \frac{d}{dz}\alpha(z) \quad \text{for} \quad \alpha \in \mathbf{h}_*^{\perp}$$

$$[L(-1), Y_*(a,z)] = \frac{d}{dz} Y_*(a,z) \quad \text{for} \quad a \in \hat{L} \tag{9.44}$$

$$Y_*(L(-1)v, z) = \frac{d}{dz} Y_*(v,z) \quad \text{for} \quad v \in V_L \tag{9.45}$$

and the definition of the vertex operators (3.40) (cf. Proposition 5.7). (Of course, the Virasoro algebra relations are not needed for this argument, and they can be proved easily after the Jacobi identity is established.) Of course, by Propositions 7.16 and 9.18 and Remark 9.16, one could alternatively establish the generalized rationality and commutativity for just two vertex operators together with the $L(-1)$ and $L(0)$-bracket properties ((6.20), (6.21)), but since the proof of generalized rationality and commutativity for more than two operators is essentially the same as the proof for two operators, the second approach would require extra work to verify (6.21).

Remark 9.20 (cf. Remark 7.19): We see that the entire discussion of duality, through Proposition 9.18, applies with the obvious modifications to modules, even in the more general sense of the modules $(W, Y, S, (\cdot,\cdot))$ described in Remark 9.7, except of course for those results which use the vacuum vector $\mathbf{1}$. In particular, in the definition of "module," the generalized Jacobi identity may be replaced by generalized rationality, commutativity and associativity.

Remark 9.21: Recall the notion of vertex operator superalgebra from Remark 6.17. In the definition of generalized vertex algebra, the case $T = 2$, $G = \mathbf{Z}/2\mathbf{Z}$ (so that $c(\cdot,\cdot)$ is trivial) and $(g,h) = gh \in \mathbf{Z}/2\mathbf{Z}$ gives a notion of *vertex superalgebra* $(V, Y, \mathbf{1}, \omega, \mathbf{Z}/2\mathbf{Z})$ (cf. Remark 9.1). Then a vertex operator superalgebra is exactly a vertex superalgebra satisfying the two grading restrictions (6.4) and (6.5). The special case of Theorem 9.8 with all the indicated restrictions on T, G, $(\cdot,\cdot)$ and $c(\cdot,\cdot)$ yields examples of vertex superalgebras Ω_*. See also Remark 12.38 below.

Chapter 10

Tensor products

In this brief chapter, we define the notion of tensor product of finitely many generalized vertex algebras, and we prove using duality that this structure in fact satisfies the axioms for a generalized vertex algebra. Given a module for each of the algebras, we analogously define the notion of the tensor product module for the tensor product algebra, and we show that it is in fact a module by using the same duality argument. The definitions and results in this chapter are modeled on the corresponding definitions and results given in [FHL] for the special case of a vertex operator algebra.

We shall define the tensor product of finitely many generalized vertex algebras

$$V_i = (V_i, Y_i, \mathbf{1}_i, \omega_i, T_i, G_i, (\cdot,\cdot)_i, c_i(\cdot,\cdot)), \qquad i = 1, ..., n. \tag{10.1}$$

Set

$$T = \text{the least common multiple of} \quad T_1, ..., T_n \tag{10.2}$$

and define the abelian group

$$G = G_1 \times \cdots \times G_n. \tag{10.3}$$

Then the sum $(\cdot,\cdot)$ of the $(\cdot,\cdot)_i$ is a symmetric $(\frac{1}{T}\mathbf{Z})/2\mathbf{Z}$-valued $\mathbf{Z}$-bilinear form on G and the product $c(\cdot,\cdot)$ of the $c_i(\cdot,\cdot)$ is an alternating $\mathbf{Z}$-bilinear form on G with values in the multiplicative group $\mathbf{C}^\times$. The vector space

$$V = V_1 \otimes \cdots \otimes V_n \tag{10.4}$$

has two natural gradations, compatible in the sense of (9.7):

$$V = \coprod_{n \in \frac{1}{T}\mathbf{Z}} V_n = \coprod_{g \in G} V^g. \tag{10.5}$$

We set

$$(10.6) \quad Y(v_1 \otimes \cdots \otimes v_n, z) = Y_1(v_1, z) \otimes \cdots \otimes Y_n(v_n, z) \quad (v_i \in V_i)$$

$$(10.7) \quad \mathbf{1} = \mathbf{1}_1 \otimes \cdots \otimes \mathbf{1}_n$$

$$(10.8) \quad \omega = \omega_1 \otimes \mathbf{1}_2 \otimes \cdots \otimes \mathbf{1}_n + \cdots + \mathbf{1}_1 \otimes \cdots \otimes \mathbf{1}_{n-1} \otimes \omega_n.$$

Then V is a tensor product of Virasoro algebra modules. All the axioms except the Jacobi identity are easy to check for the tensor product structure, and using the duality result Proposition 9.13 (with Proposition 7.11), we conclude:

Proposition 10.1 *The tensor product* $V = (V, Y, \mathbf{1}, \omega, T, G, (\cdot,\cdot), c(\cdot,\cdot))$ *is a generalized vertex algebra with rank equal to the sum of the ranks of the* V_i. □

(Note that in concluding the Jacobi identity from generalized commutativity and associativity, we are using the fact that the tensor product of the dual spaces of possibly infinite-dimensional vector spaces, such as the homogeneous components $(V_i)_n$, is dense in the dual of the tensor product.)

Given modules $(W_1, Y_1, S_1, (\cdot,\cdot)_1), \ldots, (W_n, Y_n, S_n, (\cdot,\cdot)_n)$ for $V_1, \ldots, V_n$, respectively (see Remark 9.7), using the obvious analogues of (10.3), (10.4) and (10.6) together with Remark 9.20, we have:

Proposition 10.2 *The tensor product of modules for finitely many generalized vertex algebras is a module for the tensor product algebra.* □

Remark 10.3: The tensor product of a vertex algebra and a vertex superalgebra is a vertex superalgebra (recall Remarks 9.4 and 9.21).

Chapter 11

Intertwining operators

So far, we have focused on operators parametrized by algebra elements, operators which preserve each module. As in ordinary vertex operator algebra theory (see for example [FHL]), it is natural to consider "intertwining operators" parametrized by the vectors in a module, mapping one module to another. In this chapter, we discuss duality for intertwining operators associated with module elements against vertex operators associated with algebra elements.

First we define the vector space of intertwining operators for a given triple of modules for a generalized vertex algebra. For this purpose, it turns out to be convenient and appropriate to use the variant of the notion of module based on a G-set; see Remark 9.7. As we explain after we give the general definition, in the case in which the generalized vertex algebra is an ordinary vertex algebra (or vertex operator algebra), it is still reasonable to equip modules with G-set structure in defining the notion of intertwining operator, even though in this case the group G is trivial; the resulting auxiliary structure turns out to be nonvacuous and convenient. Fusion rules are defined to be dimensions of spaces of intertwining operators. The basic duality properties (generalized rationality, commutativity and associativity) are established for pairs of operators of which one is a vertex operator parametrized by the given algebra and the second is an intertwining operator relating three given modules. As an application of such duality, we establish a result, useful even in the special case of ordinary vertex algebras, asserting that in a setting involving irreducible modules, an intertwining operator $\mathcal{Y}$ is uniquely determined by $\mathcal{Y}(w_1, z)w_2$, where w_1 and w_2 are fixed nonzero elements homogeneous with respect to the G-set grading (this last homogeneity assumption being unnecessary in the case of vertex algebras).

In order to define intertwining operators, we begin with the following data and conditions: Let $V = (V, Y, \mathbf{1}, \omega, T, G, (\cdot,\cdot), c(\cdot,\cdot))$ be a generalized vertex algebra (see Chapter 9). Let $(W_i, Y_i, S_i, (\cdot,\cdot)_i)$, $(W_j, Y_j, S_j, (\cdot,\cdot)_j)$ and $(W_k, Y_k, S_k, (\cdot,\cdot)_k)$ be V-modules (not necessarily distinct, and possibly equal to V), in the general sense given in Remark 9.7. Suppose that we are given an operation (depending on i, j and k)

$$\begin{array}{rcl} S_j \times S_k & \rightarrow & S_i \\ (s_1, s_2) & \mapsto & s_1 + s_2 \end{array} \tag{11.1}$$

which is compatible with the actions of G in the sense that

$$g + (s_1 + s_2) = s_1 + (g + s_2) = (g + s_1) + s_2$$

and which satisfies the condition

$$(g, s_1 + s_2)_i = (g, s_1)_j + (g, s_2)_k \tag{11.2}$$

for $g \in G$, $s_1 \in S_j$ and $s_2 \in S_k$. Suppose also that we have a $\mathbf{Q}/\mathbf{Z}$-valued function $(\cdot,\cdot)_{ijk}$ (depending on i, j and k) on $S_j \times S_k$ such that

$$(g + s_1, h + s_2)_{ijk} = (g, h) + (g, s_2)_k + (h, s_1)_j + (s_1, s_2)_{ijk} \tag{11.3}$$

for $g, h \in G$, $s_1 \in S_j$ and $s_2 \in S_k$, where the terms on the right-hand side are all understood to be reduced mod $\mathbf{Z}$. Suppose in addition that we are given a function

$$\begin{array}{rcl} G \times S_j & \rightarrow & \mathbf{C}^{\times} \\ (g, s) & \mapsto & c_j(g, s) \end{array} \tag{11.4}$$

which is biadditive in the sense of (9.3) and (9.4). (As before, some of these conditions could be modified; for instance, when intertwining operators are composed, as below, it is appropriate to lift the function $(\cdot,\cdot)_{ijk}$ to a $\mathbf{Q}/2\mathbf{Z}$-valued function.)

An *intertwining operator of type* $\left[\begin{smallmatrix} i \\ j\ k \end{smallmatrix}\right]$ (or $\left[\begin{smallmatrix} W_i \\ W_j\ W_k \end{smallmatrix}\right]$) associated with the given data is a linear map from $W_j \otimes W_k$ to the vector space of formal series in rational powers of z with coefficients in W_i, or equivalently, a linear map

$$\begin{array}{lcl} W_j & \to & (\mathrm{Hom}(W_k, W_i))\{z\} \\ w & \mapsto & \mathcal{Y}(w, z) = \sum_{n \in \mathbf{Q}} w_n z^{-n-1} \quad (w_n \in \mathrm{Hom}(W_k, W_i)) \end{array} \tag{11.5}$$

(recall the notation (3.42), although here we are restricting to rational powers of z) such that "all the defining properties of a module action that make sense hold." That is, for $s_1 \in S_j$, $s_2 \in S_k$, $w_1 \in W_j$, $w_2 \in W_k$, $g \in G$, $v \in V$ and $l \in \mathbf{Q}$,

$$(w_1)_l W_k^{s_2} \subset W_i^{s_1+s_2} \quad \text{if} \quad w_1 \in W_j^{s_1}; \tag{11.6}$$

$$(w_1)_l w_2 = 0 \quad \text{if } l \text{ is sufficiently large;} \tag{11.7}$$

$$\mathcal{Y}(w_1, z)|_{W_k^{s_2}} = \sum_{n \equiv (s_1, s_2)_{ijk} \bmod \mathbf{Z}} (w_1)_n z^{-n-1} \quad \text{if} \quad w_1 \in W_j^{s_1}; \tag{11.8}$$

the following (generalized) Jacobi identity holds on W_k for the operators $Y(v, z_1)$ and $\mathcal{Y}(w_1, z_2)$:

$$\begin{aligned} & z_0^{-1} \left(\frac{z_1 - z_2}{z_0}\right)^{(g,s_1)_j} \delta\left(\frac{z_1 - z_2}{z_0}\right) Y_i(v, z_1)\mathcal{Y}(w_1, z_2) \\ & - c_j(g, s_1) z_0^{-1} \left(\frac{z_2 - z_1}{z_0}\right)^{(g,s_1)_j} \delta\left(\frac{z_2 - z_1}{-z_0}\right) \mathcal{Y}(w_1, z_2) Y_k(v, z_1) \\ & = z_2^{-1} \delta\left(\frac{z_1 - z_0}{z_2}\right) \mathcal{Y}(Y_j(v, z_0)w_1, z_2) \left(\frac{z_1 - z_0}{z_2}\right)^{-g} \end{aligned} \tag{11.9}$$

if $v \in V^g$, $w_1 \in W_j^{s_1}$ (where the last factor on the right has the obvious meaning);

$$\frac{d}{dz}\mathcal{Y}(w_1, z) = \mathcal{Y}(L(-1)w_1, z), \tag{11.10}$$

where $L(-1)$ is the operator acting on W_j. For brevity, we shall sometimes write Y for Y_i, Y_j and Y_k. We denote the intertwining operator by

$$\mathcal{Y} = (\mathcal{Y}, +, (\cdot, \cdot)_{ijk}, c_j(\cdot, \cdot)). \tag{11.11}$$

Remark 11.1: Note that $Y(\cdot, z)$ acting on V is an example of an intertwining operator of type $\left[\begin{smallmatrix} V \\ V\ V \end{smallmatrix}\right]$, and $Y(\cdot, z)$ acting on a V-module W is an example of an intertwining operator of type $\left[\begin{smallmatrix} W \\ V\ W \end{smallmatrix}\right]$.

The intertwining operators of type $\begin{bmatrix} i \\ j\ k \end{bmatrix}$ associated with fixed data clearly form a vector space, which we denote by $\mathcal{V}^i_{jk}$ or $\mathcal{V}^{W_i}_{W_j W_k}$. We set

$$N^i_{jk} = \dim \mathcal{V}^i_{jk}\ (\leq \infty). \tag{11.12}$$

These numbers are called the *fusion rules* associated with the algebra, modules and auxiliary data. Then for example, assuming that V and the V-module W are nonzero, the corresponding fusion rules are positive:

$$N^V_{VV} \geq 1, \tag{11.13}$$

$$N^W_{VW} \geq 1. \tag{11.14}$$

Remark 11.2: Formulas (6.59)-(6.63) and Remark 6.11 hold in the present context.

Remark 11.3 (cf. Remarks 6.4, 6.12 and Remark 9.4): Even in the restricted case of an ordinary vertex algebra, where $T = 1$, $G = 0$ and so on, the condition (11.8) would imply that $\mathcal{Y}(w_1, z)$ involves only integral powers of z, and it is important that intertwining operators not be so restricted (see for example [FHL]). However, (11.8) is very useful, even for an ordinary vertex operator algebra, and it is valuable to equip modules W for a vertex algebra with the additional S-structure described in Remark 6.15, in such a way that the decomposition into submodules W^s $(s \in S)$ is fine enough so that the intertwining operators that one wishes to study can be given the structure axiomatized above, including (11.8). This procedure will be used later. In case the three modules W_j, W_k, W_i are irreducible, and the corresponding sets S_j, S_k, S_i consist of just the elements s_1, s_2, s_3, respectively, it is easy to see (as in [FHL], Remark 5.4.4) that (11.8) will hold if we take

$$(s_1, s_2)_{ijk} = \mathrm{wt}\, w_1 + \mathrm{wt}\, w_2 - \mathrm{wt}\, w_3 + \mathbf{Z} \in \mathbf{Q}/\mathbf{Z} \tag{11.15}$$

where w_1, w_2 and w_3 are arbitrary nonzero homogeneous elements of W_j, W_k and W_i, respectively. In typical examples, these three weights are given mod $\mathbf{Z}$ by a quadratic form, and their alternating sum (11.15) is given by a bilinear form; cf. Remark 6.5 above and (12.1) and Remark 12.1 below.

We now discuss the duality relations between the vertex operators $Y(\cdot, z_1)$ and the intertwining operators $\mathcal{Y}(\cdot, z_2)$ of type $\begin{bmatrix} i \\ j\,k \end{bmatrix}$ defined above. Using exactly the same arguments as for generalized vertex algebras (see Propositions 9.12 and 9.13, and also Propositions 7.6, 7.8 and 7.9), we have:

Proposition 11.4 *(a)* **(generalized rationality of products)** *For $g \in G, s_1 \in S_j, s_2 \in S_k$ and $v \in V^g, w_1 \in W_j^{s_1}, w_2 \in W_k^{s_2}, w' \in W_i'$, the formal series*

$$\langle w', Y(v, z_1)\mathcal{Y}(w_1, z_2)w_2\rangle (z_1 - z_2)^{(g,s_1)_j} z_1^{(g,s_2)_k} z_2^{(s_1,s_2)_{ijk}},$$

which involves only integral powers of z_1 and z_2, lies in the image of the map ι_{12}:

$$\begin{aligned}(11.16)\qquad &\langle w', Y(v, z_1)\mathcal{Y}(w_1, z_2)w_2\rangle (z_1 - z_2)^{(g,s_1)_j} z_1^{(g,s_2)_k} z_2^{(s_1,s_2)_{ijk}} \\ &\quad = \iota_{12} f(z_1, z_2),\end{aligned}$$

where the (uniquely determined) element $f \in \mathbf{C}[z_1, z_2]_S$ is of the form

$$(11.17)\qquad f(z_1, z_2) = g(z_1, z_2)/z_1^{r_1} z_2^{r_2} (z_1 - z_2)^t$$

for some $g \in \mathbf{C}[z_1, z_2]$ and $r_1, r_2, t \in \mathbf{Z}$. The integer t depends only on v and w_1; it is independent of w_2 and w'.
(b) **(generalized commutativity)** *We also have*

$$\begin{aligned}(11.18)\qquad &c_j(g, s_1)\langle w', \mathcal{Y}(w_1, z_2)Y(v, z_1)w_2\rangle (z_2 - z_1)^{(g,s_1)_j}\cdot \\ &\quad \cdot z_1^{(g,s_2)_k} z_2^{(s_1,s_2)_{ijk}} = \iota_{21} f(z_1, z_2),\end{aligned}$$

that is,

"$Y(v, z_1)\mathcal{Y}(w_1, z_2)(z_1 - z_2)^{(g,s_1)_j} z_1^{(g,s_2)_k} z_2^{(s_1,s_2)_{ijk}}$ agrees with

$$c_j(g, s_1)\mathcal{Y}(w_1, z_2)Y(v, z_1)(z_2 - z_1)^{(g,s_1)_j} z_1^{(g,s_2)_k} z_2^{(s_1,s_2)_{ijk}}$$

as operator-valued rational functions." □

Proposition 11.5 *(a)* **(generalized rationality of iterates)** *With the same notation, the formal series*

$$\langle w', \mathcal{Y}(Y(v, z_0)w_1, z_2)w_2\rangle (z_2 + z_0)^{(g,s_2)_k} z_2^{(s_1,s_2)_{ijk}} z_0^{(g,s_1)_j},$$

which involves only integral powers of z_0 and z_2, lies in the image of the map ι_{20}:

$$(11.19)\quad \begin{aligned}&\langle w', \mathcal{Y}(Y(v,z_0)w_1,z_2)w_2\rangle (z_2+z_0)^{(g,s_2)_k} z_2^{(s_1,s_2)_{ijk}} z_0^{(g,s_1)_j}\\ &\quad = \iota_{20} f(z_0+z_2,z_2),\end{aligned}$$

where f is given by (11.17), and the power r_1 of z_1 in (11.17) depends only on v and w_2; it is independent of w_1 and w'.
(b) **(generalized associativity)** *In particular,*

$$(11.20)\quad \begin{aligned}&\iota_{12}^{-1}\left(\langle w', Y(v,z_1)\mathcal{Y}(w_1,z_2)w_2\rangle (z_1-z_2)^{(g,s_1)_j} z_1^{(g,s_2)_k} z_2^{(s_1,s_2)_{ijk}}\right)\\ &\quad = \iota_{20}^{-1}\left(\langle w', \mathcal{Y}(Y(v,z_0)w_1,z_2)w_2\rangle (z_2+z_0)^{(g,s_2)_k}\cdot\right.\\ &\quad\quad \left.\cdot z_2^{(s_1,s_2)_{ijk}} z_0^{(g,s_1)_j}\right)\Big|_{z_0=z_1-z_2}.\end{aligned}$$

That is,

$$\text{“}Y(v,z_1)\mathcal{Y}(w_1,z_2)(z_1-z_2)^{(g,s_1)_j} z_1^{(g,s_2)_k} z_2^{(s_1,s_2)_{ijk}} \quad \text{agrees with}$$

$$\mathcal{Y}(Y(v,z_1-z_2)w_1,z_2)(z_1-z_2)^{(g,s_1)_j} z_1^{(g,s_2)_k} z_2^{(s_1,s_2)_{ijk}}$$

as operator-valued rational functions.” □

Remark 11.6: The analogues of Corollary 7.10 and of Proposition 7.11 hold. In particular, in the definition of the notion of intertwining operator, the Jacobi identity can be replaced by the assertions of Propositions 11.4 and 11.5.

Remark 11.7: Compositions of two or more intertwining operators are another story entirely, and the results above will not carry over in general. In the next chapter, we shall study a special situation.

Remark 11.8: In the special case of an ordinary vertex algebra equipped with the S-structure for modules discussed in Remark 11.3, the “extra factors” $z_0^{(g,s_1)_j}$ and $z_1^{(g,s_2)_k}$ in Propositions 11.4 and 11.5 drop out, and only $z_2^{(s_1,s_2)_{ijk}}$ is needed; cf. Remark 5.4.4 of [FHL].

As an application of duality, we shall prove the following:

Proposition 11.9 *With the same notation, assume that the V-modules W_j and W_k are irreducible. Fix a nonzero S_j-homogeneous vector $w_1 \in W_j^{s_1}$ and a nonzero S_k-homogeneous vector $w_2 \in W_k^{s_2}$ ($s_1 \in S_j, s_2 \in S_k$). If $\mathcal{Y}(w_1, z)w_2 = 0$, then $\mathcal{Y}(\cdot, z) = 0$, that is, $\mathcal{Y}(w, z) = 0$ for all $w \in W_j$. More generally, an intertwining operator $\mathcal{Y}^1$ of the same type as $\mathcal{Y}$ is uniquely determined by the knowledge of $\mathcal{Y}^1(w_1, z)w_2$ or in particular, by the knowledge of $\mathcal{Y}^1(w_1, z)$ or $\mathcal{Y}^1(\cdot, z)w_2$.*

Proof It is sufficient to prove the first assertion. We first show that

$$\mathcal{Y}(w_1, z) = 0. \tag{11.21}$$

For this purpose we observe that for any irreducible V-module W,

$$\begin{aligned} W = \operatorname{span}\Big\{ v_{n_1}^{(1)} \cdots v_{n_m}^{(m)} \cdot w \,|\, v^{(q)} \in V^{h_q} \\ \text{where } h_q \in G,\ n_q \in \mathbf{Q} \Big\}, \end{aligned} \tag{11.22}$$

where w is an arbitrary nonzero element of W. By Proposition 11.4 we have

$$\begin{aligned} \langle w', Y(v, z_1)\mathcal{Y}(w_1, z_2)w_2\rangle (z_1 - z_2)^{(g,s_1)_j} z_1^{(g,s_2)_k} z_2^{(s_1,s_2)_{ijk}} &= \iota_{12} f(z_1, z_2) \\ c_j(g, s_1)\langle w', \mathcal{Y}(w_1, z_2)Y(v, z_1)w_2\rangle (z_2 - z_1)^{(g,s_1)_j} \cdot \\ \cdot z_1^{(g,s_2)_k} z_2^{(s_1,s_2)_k} &= \iota_{21} f(z_1, z_2) \end{aligned}$$

for $v \in V^g$ ($g \in G$) and $w' \in W_i'$. By the assumption that $\mathcal{Y}(w_1, z_2)w_2 = 0$ we see that $\iota_{12} f(z_1, z_2) = 0$, and since ι_{12} is injective, $f(z_1, z_2) = 0$. Thus $\iota_{21} f(z_1, z_2) = 0$, which implies that

$$\langle w', \mathcal{Y}(w_1, z_2)Y(v, z_1)w_2\rangle = 0.$$

Since $w' \in W_i'$ is arbitrary we conclude that

$$\mathcal{Y}(w_1, z_2)Y(v, z_1)w_2 = 0.$$

By (11.22) and induction on m we get (11.21).

An analogous argument using (11.20) (or an application of the relevant version of the generalized associator relation (7.11)) gives

$$\mathcal{Y}(Y(v, z_0)w_1, z_2) = 0,$$

and again by (11.22) and induction we see that $\mathcal{Y}(w, z) = 0$ for all $w \in W_j$. The assertions about $\mathcal{Y}^1$ follow. □

Remark 11.10: If V is a vertex algebra (see Remark 9.4) equipped with S-structure for modules, as in Remark 11.3, then every irreducible module is clearly homogeneous with respect to the associated set S. In this context, Proposition 11.9 of course holds for any nonzero elements $w_1 \in W_j$ and $w_2 \in W_k$. Proposition 11.9 is a sharpening of Remark 5.4.6 of [FHL].

Chapter 12

Abelian intertwining algebras, third cohomology and duality

In this chapter, starting from a pair of lattices $L_0 \subset L$ with L_0 even and $\langle L_0, L \rangle \subset \mathbf{Z}$, we construct certain nontrivial intertwining operators for the vertex algebra V_{L_0}, acting on V_L, and we compute the corresponding fusion rules and the corresponding fusion algebra. We also combine these intertwining operators into a structure similar to, but not quite, a generalized vertex algebra, and we abstract and study the properties of this structure of "abelian intertwining algebra." Third cohomology of the finite abelian group L/L_0 enters intrinsically. At the end of this chapter, we mention the important special case of vertex superalgebras based canonically on integral lattices.

The notion of abelian intertwining algebra generalizes that of generalized vertex algebra (Chapter 9), which in turns generalizes that of generalized vertex operator algebra (Chapter 6) and hence that of vertex operator algebra. The axiom system for this most general structure that we consider naturally places the monodromy-related ingredients of the Jacobi identity into the context of the "cohomology theory of abelian groups" introduced by Eilenberg-Mac Lane ([E], [ML], [EML1], [EML2]; see also [JS]); M. Tierney originally directed our attention to this cohomology theory, which is not the same as ordinary group cohomology theory. An abelian intertwining algebra is based on a 3-cocycle, in the sense of this cohomology theory, for the grading group G. The Jacobi identity for an abelian intertwining algebra incorporates various pieces of data, and the 3-cocycle describes all this information in a natural way. The cohomology theory of abelian groups associates a canonical quadratic form to a 3-cohomology class, and it turns out that this quadratic form exactly defines the braid group representations attached to the algebra. The general notion of abelian intertwining algebra is also very closely related to the special class of conformal field theories studied in Appendix E of [MS], where essentially this same cohomology structure appears (even if not by name) and is treated in terms of a special case of the pentagon and hexagon identities related to Mac Lane's coherence theorem. Our notations F, Ω and B are chosen to be compatible with those in [MS].

An abelian intertwining algebra may be thought of (approximately) as being formed by adjoining to a vertex algebra the direct sum of a family of modules of a special type; there is a family of intertwining operators among these modules possessing certain generalized commutativity properties, and these intertwining operators form the basic vertex operator $Y(\cdot, z)$ which defines the "multiplication" in the algebra. As before, the algebra is graded by an abelian group G. The subspace corresponding to the identity element of G is essentially the underlying vertex algebra and the subspaces corresponding to the various group elements are the modules. The generalized commutativity naturally gives rise to one-dimensional braid group representations. These are the main reasons why we have chosen the term "abelian intertwining algebra."

In this chapter, we begin by constructing a basic example of what will be an abelian intertwining algebra by starting from the pair of lattices $L_0 \subset L$, satisfying suitable hypotheses, and by modifying the existing vertex operators $Y(\cdot, z)$ in order to create a family of intertwining operators. We then show that the resulting intertwining operators span all the possible intertwining operators, and we thereby determine the fusion rules, which are all zero and one. The "fusion algebra" is correspondingly the commutative associative group algebra $\mathbf{C}[L/L_0]$. Then we compute the Jacobi identity which results when the constructed intertwining operators are combined. We find that the analogues of the symmetric bilinear form and the alternating bilinear form on G in the definition of generalized vertex algebra are no longer bilinear, and in addition we find a new ingredient in the right-hand side of the identity: a 3-cocycle (in the sense of ordinary group cohomology theory) on the abelian group L/L_0; this cocycle essentially arises from the failures of bilinearity of the other two functions. It is known that 3-cohomology classes can be interpreted in terms of "crossed module" structure (cf.

[Br]), and we digress to compute the specific crossed module which (up to equivalence) is associated with our 3-cohomology class. Then we carry over the duality and monodromy arguments and results of Chapters 7 and 8 to the present situation, emphasizing the role of the 3-cocycle, which gives rise to a canonical quadratic form on G. The observation that a natural renormalization procedure for the intertwining operators leads to a certain change of form of the Jacobi identity helps to motivate us to formulate our structure in terms of the cohomology theory of abelian groups rather than ordinary group cohomology theory.

As a kind of summary of all these constructions and ideas, we then present our formal definition of abelian intertwining algebra, based on a normalized abelian 3-cocycle, canonically determined up to an abelian 3-coboundary when the resulting vertex operators are renormalized. We formulate a precise theorem asserting that the structure developed at the beginning of the chapter provides an example of an abelian intertwining algebra. This result will be used in Chapter 13 to construct a family of intertwining operators relating higher-level standard modules for affine Lie algebras. We give the basic properties of abelian intertwining algebras, analogous to and generalizing those of generalized vertex algebras, including the fundamental duality and monodromy results, and we specify the precise sense in which the notion of generalized vertex algebra is a special case of the notion of abelian intertwining algebra. The cohomological aspects are emphasized, including the role of the basic quadratic form on the group G which canonically determines and is determined by the abelian 3-cohomology class. This quadratic form precisely describes the one-dimensional braid group representations, which are also Hecke-algebra representations, arising in the usual way. The argument which derives generalized associativity from generalized commutativity also illustrates important aspects of the cohomological structure. Finally, we observe that the construction of tensor product algebras carries over to the present generality, with duality properties again used to prove that the tensor product structure satisfies the desired axioms.

We return to the setting of Chapter 5 with $\mathbf{h}_* = 0$, and specifically, the setting of Remark 9.10, so that L_0 is an even sublattice of the (nondegenerate) lattice L (of the same rank) such that $L \subset L_0^\circ$ and $c(\alpha, \beta) = (-1)^{\langle \alpha, \beta \rangle}$ for $\alpha, \beta \in L_0$; the positive integer q is even. Recall the coset decomposition of L with respect to L_0 given by (5.53), and in particular, the coset representatives λ_j, $j \in L/L_0$, with λ_0 chosen to be zero, as well as the coset representatives Λ_j of $\hat{L}_0$ in $\hat{L}$ (see (5.54)).

We know that $(V(0), Y, \mathbf{1}, \omega)$ is a vertex algebra and that $(V(j), Y)$ is an irreducible module, for $j \in L/L_0$. We shall discuss the intertwining operators and fusion rules for the vertex algebra $V(0)$ and its modules $V(j)$; these operators are defined in (11.5)-(11.12), with conditions (11.6) and (11.8) omitted and with (11.9) replaced by the ordinary Jacobi identity (recall Remark 11.3).

It is important to keep in mind that for the results in this chapter, including the construction of an "abelian intertwining algebra" based on the pair L, L_0, it will *not* be necessary for L to the full dual lattice of the even lattice L_0. Such pairs do of course provide natural families of examples. The important case of the pair P, Q, where P is the weight lattice of a simple Lie algebra of type A, D or E and Q is the root lattice will be considered in Chapter 13.

However, it is natural to work in the more refined setting suggested by Remark 11.3. Accordingly, we assign to each module $V(j)$ (including $V(0)$) a set S_j consisting of the single element $j \in L/L_0$, and we take the various operations $+$ among the sets S_j to be those induced by addition in the abelian group L/L_0. More precisely, for $j, k \in L/L_0$, we define the obvious operation from $S_j \times S_k$ to S_{j+k}, but we leave undefined a "plus" operation from $S_j \times S_k$ to S_i for $i \neq j+k$. In particular, S_0 is the one-element abelian group, which plays the role of the group G, and the $(\frac{1}{T}\mathbf{Z})/2\mathbf{Z}$-valued (that is, $\mathbf{Z}/2\mathbf{Z}$-valued) function $(\cdot,\cdot)$ on $G \times S_j$ (recall Remark 6.15) is of course the zero function. Intertwining operators of type $\begin{bmatrix} i \\ j\ k \end{bmatrix}$ are defined only when $i = j + k$ (later, we shall see that this is reasonable), and in this case, we define the function $(\cdot,\cdot)_{ijk}$ by:

$$
\begin{aligned}
(j,k)_{ijk} &= -\langle \lambda_j, \lambda_k \rangle + \mathbf{Z} = \frac{1}{2}\langle \lambda_j, \lambda_j \rangle + \frac{1}{2}\langle \lambda_k, \lambda_k \rangle \\
&\quad -\frac{1}{2}\langle \lambda_j + \lambda_k, \lambda_j + \lambda_k \rangle + \mathbf{Z} \\
&= \frac{1}{2}\langle \lambda_j, \lambda_j \rangle + \frac{1}{2}\langle \lambda_k, \lambda_k \rangle - \frac{1}{2}\langle \lambda_i, \lambda_i \rangle + \mathbf{Z} \in \mathbf{Q}/\mathbf{Z}.
\end{aligned}
\tag{12.1}
$$

Note that this is independent of the choice of the coset representatives λ_j and that we are using the facts that $\lambda_j + \lambda_k - \lambda_i$ lies in the even lattice L_0 and that $\langle L_0, L \rangle \subset \mathbf{Z}$. Compare this with (11.15) and Remark 6.5. Later we shall abstract this structure in terms of a quadratic form and its associated bilinear form; see (12.88), (12.103), (12.114) and Remarks 12.22, 12.28, 12.29 and 12.36 below. The function $c_j(\cdot,\cdot)$ from $G \times S_j$ to $\mathbf{C}^\times$ is of course the trivial function with value 1.

Remark 12.1: For $j, k \in L/L_0$, the intertwining operators of type $\begin{bmatrix} j+k \\ j\ k \end{bmatrix}$ as originally defined above (that is, with conditions (11.6) and (11.8) omitted and with (11.9) replaced by the ordinary Jacobi identity) agree precisely with the intertwining operators of type $\begin{bmatrix} j+k \\ j\ k \end{bmatrix}$ in the new sense (recall (11.15)). We shall see below that for $i \neq j+k$, there are no nonzero intertwining operators of type $\begin{bmatrix} i \\ j\ k \end{bmatrix}$ in the original sense, and we have not defined any in the new sense, so that again the old and new definitions agree.

In order to construct intertwining operators, we proceed as follows: For $\alpha \in L$, we define operators $e^{i\pi\alpha}$ and $c(\cdot, \alpha)$ on V_L by

$$e^{i\pi\alpha} \cdot u \otimes \iota(b) = e^{i\pi\langle\alpha,\bar{b}\rangle} u \otimes \iota(b)$$
(12.2)
$$c(\cdot, \alpha) \cdot u \otimes \iota(b) = c(\bar{b}, \alpha) u \otimes \iota(b)$$

($u \in M(1)$ and $b \in \hat{L}$). Note that if $\alpha, \bar{b} \in L_0$, these two operators agree. For $w \in V(j)$ ($j \in L/L_0$) we define an operator

$$\mathcal{Y}_{\lambda_j}(w, z) = Y(w, z) e^{i\pi\lambda_j} c(\cdot, \lambda_j) \tag{12.3}$$

on V_L. Note that $\mathcal{Y}_{\lambda_0} = Y$ since $\lambda_0 = 0$.

Proposition 12.2 *For $k \in L/L_0$, the restriction of the operator $\mathcal{Y}_{\lambda_j}(\cdot, z)$ to $V(k)$ is a nonzero intertwining operator of type $\begin{bmatrix} i \\ j\ \ k \end{bmatrix}$, where $i = j + k$.*

Proof It is easy to see that $\mathcal{Y}_{\lambda_j}(w, z)V(k) \subset V(i)\{z\}$ (with rational powers of z) for $w \in V(j)$ and that the operator is nonzero. Since $\mathcal{Y}_{\lambda_j}(\cdot, z)$ is obtained from $Y(\cdot, z)$ by multiplying by an operator which is independent of z and since $L(-1)$ preserves $V(j)$, we see from Proposition 5.7 that

$$\mathcal{Y}_{\lambda_j}(L(-1)w, z) = \frac{d}{dz}\mathcal{Y}_{\lambda_j}(w, z) \quad \text{for} \quad w \in V(j). \tag{12.4}$$

The truncation property (11.7) being obvious, it remains to prove the (ordinary) Jacobi identity for the operators $Y(u, z_1)$ and $\mathcal{Y}_{\lambda_j}(w, z_2)$ for

$u \in V(0)$ and $w \in V(j)$. Let $u^*, w^* \in M(1)$ and $a \in \hat{L}_0$, $b \in \hat{L}_0\Lambda_j$ (recall (5.54)), and consider the elements

$$u = u^* \otimes \iota(a) \in V(0), \quad w = w^* \otimes \iota(b) \in V(j).$$

By Theorem 5.1 (with $\mathbf{h}_* = 0$ and $\langle \bar{a}, L\rangle \subset \mathbf{Z}$), or Theorem 8.8.23 of [FLM3], we have

$$\begin{aligned} &z_0^{-1}\delta\left(\frac{z_1 - z_2}{z_0}\right) Y(u, z_1)Y(w, z_2) \\ &- (-1)^{\langle \bar{a}, \bar{b}\rangle} c(\bar{a}, \bar{b}) z_0^{-1}\delta\left(\frac{z_2 - z_1}{-z_0}\right) Y(w, z_2)Y(u, z_1) \\ &= z_2^{-1}\delta\left(\frac{z_1 - z_0}{z_2}\right) Y(Y(u, z_0)w, z_2). \end{aligned} \tag{12.5}$$

Multiplying (12.5) by $e^{i\pi\lambda_j}c(\cdot, \lambda_j)$ on the right and noting that for the operator $a \in \hat{L}_0$ occurring as a factor of $Y(u, z)$ and for $\lambda \in L$,

$$a e^{i\pi\lambda} c(\cdot, \lambda) = (-1)^{\langle \lambda, \bar{a}\rangle} c(\lambda, \bar{a}) e^{i\pi\lambda} c(\cdot, \lambda) a \tag{12.6}$$

(since $\langle \bar{a}, L\rangle \subset \mathbf{Z}$), and that

$$(-1)^{\langle \bar{a}, \bar{b}\rangle} c(\bar{a}, \bar{b}) = (-1)^{\langle \bar{a}, \lambda_j\rangle} c(\bar{a}, \lambda_j), \tag{12.7}$$

we get the desired identity

$$\begin{aligned} &z_0^{-1}\delta\left(\frac{z_1 - z_2}{z_0}\right) Y(u, z_1)\mathcal{Y}_{\lambda_j}(w, z_2) \\ &-z_0^{-1}\delta\left(\frac{z_2 - z_1}{-z_0}\right) \mathcal{Y}_{\lambda_j}(w, z_2)Y(u, z_1) \\ &= z_2^{-1}\delta\left(\frac{z_1 - z_0}{z_2}\right) \mathcal{Y}_{\lambda_j}(Y(u, z_0)w, z_2). \quad \square \end{aligned} \tag{12.8}$$

Although the intertwining operator $\mathcal{Y}_{\lambda_j}(\cdot, z)|_{V(k)}$ depends on the choice of the representative λ_j, we next show that different representatives give the same intertwining operator up to a scalar multiple:

Proposition 12.3 *We have*

$$\mathcal{Y}_{\lambda_j+\alpha}(\cdot, z)|_{V(k)} = (-1)^{\langle \alpha, \lambda_k\rangle} c(\lambda_k, \alpha) \mathcal{Y}_{\lambda_j}(\cdot, z)|_{V(k)} \tag{12.9}$$

for $\alpha \in L_0$. Furthermore, the factor $(-1)^{\langle \alpha, \lambda_k\rangle} c(\lambda_k, \alpha)$ depends only on α and k, that is,

$$(-1)^{\langle \alpha, \lambda_k + \beta\rangle} c(\lambda_k + \beta, \alpha) = (-1)^{\langle \alpha, \lambda_k\rangle} c(\lambda_k, \alpha)$$

for $\beta \in L_0$.

Proof Use the fact that

$$e^{i\pi(\lambda_j+\alpha)}c(\cdot,\lambda_j+\alpha)\cdot\iota(b\Lambda_k) = (-1)^{\langle\alpha,\lambda_k\rangle}c(\lambda_k,\alpha)e^{i\pi\lambda_j}c(\cdot,\lambda_j)\iota(b\Lambda_k)$$

for $b \in \hat{L}_0$ and the definition (12.3). □

Next we shall prove a stronger statement – that any intertwining operator of type $\begin{bmatrix} & i & \\ j & & k \end{bmatrix}$ with $i = j + k$ is a scalar multiple of $\mathcal{Y}_{\lambda_j}(\cdot, z)|_{V(k)}$.

Let $i, j, k \in L/L_0$ and let

$$\mathcal{Y} : V(j) \otimes V(k) \to V(i)\{z\}$$

be an arbitrary intertwining operator of type $\begin{bmatrix} & i & \\ j & & k \end{bmatrix}$. Then the fact that

(12.10) $\quad [u_0, \mathcal{Y}(w, z)] = \mathcal{Y}(u_0 w, z) \quad$ for $\quad u \in V(0),\ w \in V(j)$

(which follows by applying $\mathrm{Res}_{z_0}\mathrm{Res}_{z_1}$ to the Jacobi identity (12.8)) implies that if w is a simultaneous eigenvector for the action of $\mathbf{h}$, then $\mathcal{Y}(w, z)$ permutes $\mathbf{h}$-eigenvectors correspondingly. Thus from (3.17) we have:

Lemma 12.4 *For $a \in \hat{L}_0\Lambda_j$ and $b \in \hat{L}_0\Lambda_k$,*

$$\mathcal{Y}(a, z)\iota(b) \subset (M(1) \otimes \iota(ab))\{z\}. \qquad \square$$

This shows that

$$\mathcal{Y}(a, z)\iota(b) \subset V(j + k)\{z\},$$

in other words, if $i \neq j + k$, then $\mathcal{Y}(a, z)\iota(b) = 0$ for all $a \in \hat{L}_0\Lambda_j$ and $b \in \hat{L}_0\Lambda_k$. By Proposition 11.9 and Remark 11.10, $\mathcal{Y}(\cdot, z) = 0$ in this case:

Proposition 12.5 *If $i \neq j + k$, $\mathcal{V}^i_{jk} = 0$, that is, $N^i_{jk} = 0$.* □

Now we assume that $i = j + k$. Then (by Remark 12.1) $\mathcal{Y}(\iota(\Lambda_j), z)$ has an expansion (as in Remark 5.4.4 of [FHL])

(12.11) $$\mathcal{Y}(\iota(\Lambda_j), z) = \sum_{n\in\mathbf{Z}} \mathcal{Y}_n(\iota(\Lambda_j))z^{-n+\langle\lambda_j,\lambda_k\rangle}$$

(where we use the fact that $\mathrm{wt}\,\iota(a) = \frac{1}{2}\langle\bar{a},\bar{a}\rangle$ for $a \in \hat{L}$). Then from the Jacobi identity (12.8) and (12.4) we have the following commutation relations between $L(0)$ and $\mathcal{Y}_n(\iota(\Lambda_j))$:

$$(12.12)\quad [L(0), \mathcal{Y}_n(\iota(\Lambda_j))] = \left(\frac{1}{2}\langle\lambda_j,\lambda_j\rangle + \langle\lambda_j,\lambda_k\rangle - n\right)\mathcal{Y}_n(\iota(\Lambda_j)).$$

Lemma 12.6 *If* $n > 0$, $\mathcal{Y}_n(\iota(\Lambda_j))\iota(\Lambda_k) = 0$, *and* $\mathcal{Y}_0(\iota(\Lambda_j))\iota(\Lambda_k) \subset \mathbf{C}\iota(\Lambda_j\Lambda_k)$.

Proof Note that $\iota(\Lambda_j\Lambda_k)$ has the lowest possible weight in $M(1) \otimes \iota(\Lambda_j\Lambda_k)$ and that it is unique (up to scalar multiple) with this property. By (12.12),

$$\mathrm{wt}\,(\mathcal{Y}_n(\iota(\Lambda_j))\iota(\Lambda_k)) = \frac{1}{2}\langle\lambda_j+\lambda_k,\lambda_j+\lambda_k\rangle - n,$$

and by Lemma 12.4, $\mathcal{Y}_n(\iota(\Lambda_j))\iota(\Lambda_k) \in M(1) \otimes \iota(\Lambda_j\Lambda_k)$. The lemma follows. □

Lemma 12.7 *If* $\mathcal{Y}_0(\iota(\Lambda_j))\iota(\Lambda_k) = 0$, *then* $\mathcal{Y}(\cdot, z) = 0$.

Proof By Proposition 11.9 and Remark 11.10, it is sufficient to prove that $\mathcal{Y}(\iota(\Lambda_j), z)\iota(\Lambda_k)$
$=0$. From Lemma 12.6 we need only show that $\mathcal{Y}_n(\iota(\Lambda_j))\iota(\Lambda_k) = 0$ for $n < 0$. We prove this by induction on n. Assume that $\mathcal{Y}_q(\iota(\Lambda_j))\iota(\Lambda_k) = 0$ for all $q > n$ $(n < 0)$. By the Jacobi identity (12.8) we have the commutation relations

$$[\alpha(m), \mathcal{Y}_n(\iota(\Lambda_j))] = \langle\alpha,\lambda_j\rangle\mathcal{Y}_{m+n}(\iota(\Lambda_j))$$

for $\alpha \in \mathbf{h}$ and $m \in \mathbf{Z}$. Thus

$$\alpha(m)\,(\mathcal{Y}_n(\iota(\Lambda_j))\iota(\Lambda_k))$$
$$= \mathcal{Y}_n(\iota(\Lambda_j))\,(\alpha(m)\iota(\Lambda_k)) + \langle\alpha,\lambda_j\rangle\mathcal{Y}_{m+n}(\iota(\Lambda_j))\iota(\Lambda_k) = 0$$

if $m > 0$, and therefore,

$$\mathcal{Y}_n(\iota(\Lambda_j))\iota(\Lambda_k) \in \mathbf{C}\iota(\Lambda_j\Lambda_k).$$

However,

$$\mathrm{wt}\,\mathcal{Y}_n(\iota(\Lambda_j))\iota(\Lambda_k) = \mathrm{wt}\,\iota(\Lambda_j\Lambda_k) - n > \mathrm{wt}\,\iota(\Lambda_j\Lambda_k),$$

so that $\mathcal{Y}_n(\iota(\Lambda_j))\iota(\Lambda_k) = 0$. □

Proposition 12.8 *We have* $N^i_{jk} \leq 1$.

Proof Let $\mathcal{Y}^1, \mathcal{Y}^2$ be two intertwining operators of type $\left[\begin{smallmatrix} & i & \\ j & & k \end{smallmatrix}\right]$ with $\mathcal{Y}^1 \neq 0$. By Lemmas 12.6 and 12.7, $\mathcal{Y}^1_0(\iota(\Lambda_j))\iota(\Lambda_k)$ is a nonzero multiple of $\iota(\Lambda_j\Lambda_k)$, and there exists a scalar c such that

$$\mathcal{Y}^2_0(\iota(\Lambda_j))\iota(\Lambda_k) = c\mathcal{Y}^1_0(\iota(\Lambda_j)\iota(\Lambda_k),$$

that is,

$$(\mathcal{Y}^2 - c\mathcal{Y}^1)_0(\iota(\Lambda_j))\iota(\Lambda_k) = 0.$$

Noting that $\mathcal{Y}^2 - c\mathcal{Y}^1$ is also an intertwining operator of the same type and applying Lemma 12.7 we conclude that

$$\mathcal{Y}^2 - c\mathcal{Y}^1 = 0,$$

i.e.,

$$\mathcal{Y}^2 = c\mathcal{Y}^1. \qquad \square$$

Proposition 12.9 *If* $i = j + k$, *the space* $\mathcal{V}^i_{jk}$ *is spanned by the intertwining operator* $\mathcal{Y}_{\lambda_j}$ *defined in (12.3). In particular,* $N^i_{jk} = 1$. $\square$

Combined with Proposition 12.5, this describes all the intertwining operators and fusion rules.

Let $(V, Y, \mathbf{1}, \omega, T, G, (\cdot,\cdot), c(\cdot,\cdot))$ be a generalized vertex algebra. Let $\mathcal{C}$ be a set of (equivalence classes of) V-modules with appropriate auxiliary data such that for $i, j, k \in \mathcal{C}$,

$$N^i_{jk} < \infty \tag{12.13}$$

and

$$\{l \in \mathcal{C} | N^l_{jk} \neq 0\} \quad \text{is a finite set.} \tag{12.14}$$

The *fusion algebra for* V *and* $\mathcal{C}$ *over* $\mathbf{C}$ is the (nonassociative in general) algebra with basis $\{e_j | j \in \mathcal{C}\}$ and defining relations (as in [V])

$$e_j \times e_k = \sum_{i \in \mathcal{C}} N^i_{jk} e_i \tag{12.15}$$

for $j, k \in \mathcal{C}$.

From Propositions 12.5 and 12.9 we immediately have:

Corollary 12.10 *The set of equivalence classes of the modules* $V(j)$ *for* $j \in L/L_0$ *satisfies the conditions (12.13) and (12.14) and the corresponding fusion algebra is the (commutative and associative) group algebra* $\mathbf{C}[L/L_0]$ *of the finite abelian group* L/L_0. □

Next we shall discuss relations among the intertwining operators $\mathcal{Y}_{\lambda_j}$ defined in (12.3). A straightforward computation using Theorem 5.1 as in the proof of Proposition 12.2 gives us a modified form of the Jacobi identity for three operators. Two formulas which enter into the computation are:

$$ae^{i\pi\lambda}c(\cdot,\lambda) = e^{-i\pi\langle\lambda,\bar{a}\rangle}c(\lambda,\bar{a})e^{i\pi\lambda}c(\cdot,\lambda)a \tag{12.16}$$

for $a \in \hat{L}$ and $\lambda \in L$ (cf. (12.6), in which $a \in \hat{L}_0$) and

$$c(\lambda_{j_2},\alpha_1)c(\alpha_1,\alpha_2)c(\alpha_2,\lambda_{j_1})c(\lambda_{j_1},\lambda_{j_2}) = (-1)^{\langle\alpha_1-\lambda_{j_1},\alpha_2-\lambda_{j_2}\rangle} \tag{12.17}$$

for $j_1, j_2 \in L/L_0$ and $\alpha_1 \in \lambda_{j_1}+L_0$, $\alpha_2 \in \lambda_{j_2}+L_0$ (that is, $\alpha_1 \in j_1$, $\alpha_2 \in j_2$). The result is (Proposition 12.3 being used for the second part):

Theorem 12.11 *Let* $j_1, j_2 \in L/L_0$ *and* $w_1 \in V(j_1)$, $w_2 \in V(j_2)$. *Then as operators on* V_L,

$$\begin{aligned}
&z_0^{-1}\left(\frac{z_1-z_2}{z_0}\right)^{-\langle\lambda_{j_1},\lambda_{j_2}\rangle}\delta\left(\frac{z_1-z_2}{z_0}\right)\mathcal{Y}_{\lambda_{j_1}}(w_1,z_1)\mathcal{Y}_{\lambda_{j_2}}(w_2,z_2)\\
&-c(\lambda_{j_2},\lambda_{j_1})z_0^{-1}\left(\frac{z_2-z_1}{z_0}\right)^{-\langle\lambda_{j_1},\lambda_{j_2}\rangle}\delta\left(\frac{z_2-z_1}{-z_0}\right)\mathcal{Y}_{\lambda_{j_2}}(w_2,z_2)\mathcal{Y}_{\lambda_{j_1}}(w_1,z_1)\\
&= z_2^{-1}\delta\left(\frac{z_1-z_0}{z_2}\right)\mathcal{Y}_{\lambda_{j_1}+\lambda_{j_2}}(\mathcal{Y}_{\lambda_{j_1}}(w_1,z_0)w_2,z_2)\left(\frac{z_1-z_0}{z_2}\right)^{\lambda_{j_1}},
\end{aligned} \tag{12.18}$$

or equivalently, as operators on $V(j_3)$, $j_3 \in L/L_0$,

$$\begin{aligned}
&z_0^{-1}\left(\frac{z_1-z_2}{z_0}\right)^{-\langle\lambda_{j_1},\lambda_{j_2}\rangle}\delta\left(\frac{z_1-z_2}{z_0}\right)\mathcal{Y}_{\lambda_{j_1}}(w_1,z_1)\mathcal{Y}_{\lambda_{j_2}}(w_2,z_2)\\
&-c(\lambda_{j_2},\lambda_{j_1})z_0^{-1}\left(\frac{z_2-z_1}{z_0}\right)^{-\langle\lambda_{j_1},\lambda_{j_2}\rangle}\delta\left(\frac{z_2-z_1}{-z_0}\right)\mathcal{Y}_{\lambda_{j_2}}(w_2,z_2)\mathcal{Y}_{\lambda_{j_1}}(w_1,z_1)\\
&= z_2^{-1}\delta\left(\frac{z_1-z_0}{z_2}\right)\mathcal{Y}_{\lambda_{j_1+j_2}}(\mathcal{Y}_{\lambda_{j_1}}(w_1,z_0)w_2,z_2)\left(\frac{z_1-z_0}{z_2}\right)^{\langle\lambda_{j_1},\lambda_{j_3}\rangle}\cdot\\
&\quad\cdot(-1)^{\langle\lambda_{j_1}+\lambda_{j_2}-\lambda_{j_1+j_2},\lambda_{j_3}\rangle}c(\lambda_{j_3},\lambda_{j_1}+\lambda_{j_2}-\lambda_{j_1+j_2}).
\end{aligned} \tag{12.19}$$

□

Remark 12.12: If we were to combine the operators $\mathcal{Y}_{\lambda_j}$, $j \in L/L_0$, into a single operator from $V_L = \coprod_{j \in L/L_0} V(j)$ to a space of formal series with coefficients in $\operatorname{End} V_L$, this operator would not quite satisfy the Jacobi identity (9.14) defining a generalized vertex algebra. Specifically, if we were to take L/L_0 as the group G in the definition of generalized vertex algebra, then we would encounter the following problems: The symmetric function $-\langle \lambda_{j_1}, \lambda_{j_2} \rangle + 2\mathbf{Z}$ from $L/L_0 \times L/L_0$ to $\mathbf{Q}/2\mathbf{Z}$ is not $\mathbf{Z}$-bilinear (although it is bilinear when reduced mod $\mathbf{Z}$); the "alternating" and skew-symmetric function $c(\lambda_{j_2}, \lambda_{j_1})$ from $L/L_0 \times L/L_0$ to $\mathbf{C}^\times$ is not $\mathbf{Z}$-bilinear; and (12.19) includes the extra factor

$$h(j_1, j_2, j_3) = (-1)^{\langle \lambda_{j_1} + \lambda_{j_2} - \lambda_{j_1+j_2}, \lambda_{j_3} \rangle} c(\lambda_{j_3}, \lambda_{j_1} + \lambda_{j_2} - \lambda_{j_1+j_2}), \tag{12.20}$$

which defines a function from $L/L_0 \times L/L_0 \times L/L_0$ to $\mathbf{C}^\times$. It is illuminating to write the initial part of the second term on the left-hand side of (12.19) as

$$- c(\lambda_{j_2}, \lambda_{j_1}) e^{-i\pi \langle \lambda_{j_1}, \lambda_{j_2} \rangle} z_0^{-1} \left(\frac{z_2 - z_1}{e^{i\pi} z_0} \right)^{-\langle \lambda_{j_1}, \lambda_{j_2} \rangle} \delta \left(\frac{z_2 - z_1}{-z_0} \right), \tag{12.21}$$

since this shows that what is important about the two failures of bilinearity mentioned above is the failure of the function

$$(j_1, j_2) \mapsto c(\lambda_{j_2}, \lambda_{j_1}) e^{-i\pi \langle \lambda_{j_1}, \lambda_{j_2} \rangle} \tag{12.22}$$

on $L/L_0 \times L/L_0$ to be bilinear. But this failure is exactly given by (12.20). It is also worth noting that in the three expressions $\left(\frac{z_1 - z_2}{z_0}\right)^{-\langle \lambda_{j_1}, \lambda_{j_2} \rangle}$, $\left(\frac{z_2 - z_1}{e^{i\pi} z_0}\right)^{-\langle \lambda_{j_1}, \lambda_{j_2} \rangle}$ and $\left(\frac{z_1 - z_0}{z_2}\right)^{\langle \lambda_{j_1}, \lambda_{j_3} \rangle}$, multiplied by their respective δ-function expressions, the choice of section

$$j \mapsto \lambda_j \tag{12.23}$$

of the extension $L \to L/L_0$ is irrelevant. The important features of this situation will be studied and abstracted in the rest of this chapter.

We next interpret the function h in terms of group cohomology theory. (See e.g. [Br] for background.) We continue to work in the main setting of this chapter – the setting of Remark 9.10. First note that the values of h lie in the cyclic group $\langle \omega_q \rangle$ generated by ω_q; recall that q is even. Also note the occurrence of the 2-cocycle

$$\begin{aligned} L/L_0 \times L/L_0 &\to L_0 \\ (j_1, j_2) &\mapsto \varepsilon(j_1, j_2) = \lambda_{j_1} + \lambda_{j_2} - \lambda_{j_1+j_2} \end{aligned} \tag{12.24}$$

in the expression for h. In fact, we may write h in the form

$$(12.25)\ h(j_1,j_2,j_3)=(-1)^{\langle\varepsilon(j_1,j_2),\lambda_{j_3}\rangle}c(\lambda_{j_3},\varepsilon(j_1,j_2))=k(\varepsilon(j_1,j_2),j_3)^{-1},$$

where k is the biadditive map

$$(12.26)\qquad \begin{array}{rrcl} k: & L_0\times L/L_0 & \to & \langle\omega_q\rangle \\ & (\alpha,\beta+L_0) & \mapsto & (-1)^{\langle\alpha,\beta\rangle}c(\alpha,\beta) \end{array}$$

($\alpha\in L_0, \beta\in L$); note that k is indeed well defined, since its lift to $L_0\times L$ vanishes on $L_0\times L_0$. Using this structure, it is straightforward to verify:

Proposition 12.13 *The function h is a 3-cocycle with values in $\langle\omega_q\rangle$, viewed as a trivial L/L_0-module:*

$$(12.27)\qquad h\in Z^3(L/L_0,\langle\omega_q\rangle).$$

Moreover, its cohomology class $[h]$ is independent of the choice of section $j\mapsto\lambda_j$; it depends only on the class $[\varepsilon]\in H^2(L/L_0,L_0)$. In particular, the correspondence $\varepsilon\mapsto h$ based on the map k defines a linear map

$$(12.28)\qquad \begin{array}{ccc} H^2(L/L_0,L_0) & \to & H^3(L/L_0,\langle\omega_q\rangle) \\ [\varepsilon] & \mapsto & [h]. \end{array}\qquad\square$$

Remark 12.14: The function k is in some sense the source of the cohomology class $[h]$ (but see also Theorem 12.15 and Remark 12.17 below, as well as the discussion later in the chapter). Recall also that this function already occurred in the Jacobi identity (12.5) (and in Theorem 8.8.23 of [FLM3]), and in fact it was precisely the desire to remove this factor and to produce the ordinary Jacobi identity (12.8) for the intertwining operator $\mathcal{Y}_{\lambda_j}$ which led us, in (12.3), to modify the operator Y by multiplying it by $e^{i\pi\lambda_j}c(\cdot,\lambda_j)$. (Compare the ordinary Jacobi identity (12.8) defining the intertwining operators with the more general identity (12.19) relating two such operators and involving the 3-cocycle h.) Note that this factor $k(\cdot,\cdot)$ is not in general removable by a choice of $\hat{L}$, i.e., a choice of alternating biadditive map $c: L\times L\to\mathbf{C}^\times$ such that $c(\alpha,\beta)=(-1)^{\langle\alpha,\beta\rangle}$ for $\alpha\in L_0$, $\beta\in L$: For instance, we can simply take L_0 to be the (rank 1) root lattice of $\mathbf{sl}(2)$ generated by α

with $\langle\alpha, \alpha\rangle = 2$ and L to be the weight lattice, generated by $\frac{\alpha}{2}$; if c existed, then we would have the condition

$$1 = c\left(\frac{\alpha}{2}, \frac{\alpha}{2}\right)^2 = c\left(\alpha, \frac{\alpha}{2}\right) = (-1)^{\langle\alpha, \frac{\alpha}{2}\rangle} = -1.$$

It is interesting to view the cohomology class $[h]$ in terms of the known interpretation of third cohomology groups in group cohomology theory via "crossed modules" ([Ger]; cf. [Br]): Suppose that we are given an action of a group E on a group N, denoted $(e, n) \mapsto {}^e n$, together with a homomorphism $\tau : N \to E$ such that

$$(12.29) \qquad {}^{\tau(n)}n' = nn'n^{-1}$$

$$(12.30) \qquad \tau({}^e n) = e\tau(n)e^{-1}$$

for $n, n' \in N$, $e \in E$. Then N (together with τ and the action) is called a *crossed module over* E. Given such a structure, we have an exact sequence

$$(12.31) \qquad 1 \to A \xrightarrow{i} N \xrightarrow{\tau} E \xrightarrow{\pi} G \to 1$$

where $A = \operatorname{Ker} \tau$ and $G = \operatorname{Coker} \tau$ (defined since $\operatorname{Im} \tau$ is normal in E) and A, which is central in N, is naturally a G-module. Given an arbitrary group G and G-module A, there is a bijection (described in [Br]; see below) between $H^3(G, A)$ and the set of (suitably defined) equivalence classes of exact sequences (12.31) where N is a crossed module over E such that the action of E on N induces the given action of G on A. Under this bijection, the zero element of $H^3(G, A)$ corresponds to those sequences (12.31) that can be derived from a single group H such that $N \lhd H$, $H/N = G$ and $H/A = E$, where H induces the given action of E on N (see e.g. [Ho]).

We shall need the precise decomposition (cf. [Br]) of the element of $H^3(G, A)$ associated with the sequence (12.31): Choose a set-theoretic section s: $G \to E$ of π, and define f: $G \times G \to \operatorname{Ker} \pi$ by

$$(12.32) \qquad s(g)s(h) = f(g, h)s(gh),$$

so that by the associativity of multiplication in E,

$$(12.33) \qquad f(g, h)f(gh, k) = {}^{s(g)}f(h, k)f(g, hk)$$

for $g, h, k \in G$, where $s(g)$ acts by left conjugation on E. Lift f to a function $F : G \times G \to N$ and define $z : G \times G \times G \to A$ by

$$(12.34) \qquad {}^{s(g)}F(h, k)F(g, hk) = i(z(g, h, k))F(g, h)F(gh, k),$$

$s(g)$ now acting on N. Then z is a 3-cocycle, and its cohomology class, independent of the choices of s and F, is the desired one.

We relate this viewpoint to our situation. Consider the central extension

$$1 \to \langle \omega_q \rangle \to \hat{L}_0 \bar{\to} L_0 \to 1 \tag{12.35}$$

of L_0 defined (up to equivalence) by the commutator condition

$$aba^{-1}b^{-1} = (-1)^{\langle \bar{a}, \bar{b} \rangle} \quad \text{for} \quad a, b \in \hat{L}_0 \tag{12.36}$$

(cf. [FLM3], Section 5.2). Form the exact sequence

$$1 \to \langle \omega_q \rangle \to \hat{L}_0 \bar{\to} L \to L/L_0 \to 1, \tag{12.37}$$

where the maps are the obvious ones, using the maps in (12.35). Consider the biadditive map

$$\begin{array}{rcl} A : L \times L_0 & \to & \langle \omega_q \rangle \\ (\alpha, \beta) & \mapsto & (-1)^{\langle \alpha, \beta \rangle} \end{array} \tag{12.38}$$

and let L act on $\hat{L}_0$ as follows:

$$\alpha \cdot b = A(\alpha, \bar{b}) b \tag{12.39}$$

for $\alpha \in L$, $b \in \hat{L}_0$. Since $A|_{L_0 \times L_0}$ agrees with the commutator map (12.36), we see that the action of L on $\hat{L}_0$ and the map $\hat{L}_0 \to L$ give $\hat{L}_0$ crossed-module structure over L, and that the induced action of L/L_0 on $\langle \omega_q \rangle$ is trivial. To this information is associated an element of $H^3(L/L_0, \langle \omega_q \rangle)$ as above, and it turns out that (under a mild assumption) this element is precisely $[h]$:

Theorem 12.15 *Assume that the even integer q is such that*

$$\langle \alpha, \beta \rangle \in \frac{2}{q}\mathbf{Z} \quad \text{for} \quad \alpha, \beta \in L. \tag{12.40}$$

Then the class $[h] \in H^3(L/L_0, \langle \omega_q \rangle)$ corresponds to the equivalence class of the exact sequence (12.37) equipped with the crossed-module structure defined by (12.38),(12.39). Moreover, $[h]$ is independent of the choice of the alternating bilinear map $c : L \times L \to \langle \omega_q \rangle$ (subject to our constraint that $c(\alpha, \beta) = (-1)^{\langle \alpha, \beta \rangle}$ for $\alpha, \beta \in L_0$).

Proof Applying the prescription above to the sequence (12.37) we use the section λ (12.23) and the corresponding cocycle ε (12.24), which we lift to a function

$$E : L/L_0 \times L/L_0 \to \hat{L}_0. \tag{12.41}$$

The 3-cocycle z defined by (12.34) becomes the function

$$C : L/L_0 \times L/L_0 \times L/L_0 \to \langle \omega_q \rangle \tag{12.42}$$

given by

$$\begin{aligned} C(j_1, j_2, j_3) = {} & A(\lambda_{j_1}, \varepsilon(j_2, j_3))E(j_2, j_3)E(j_1, j_2 + j_3)\cdot \\ & \cdot E(j_1 + j_2, j_3)^{-1}E(j_1, j_2)^{-1}, \end{aligned} \tag{12.43}$$

with A as in (12.38). However, we know that if we carry out the analogous procedure with the function $A : L \times L_0 \to \langle \omega_q \rangle$ replaced by $c|_{L \times L_0}$, where c is an arbitrary map subject to the conditions in the statement of the Theorem, i.e., the conditions in Remark 9.10, then the resulting 3-cocycle analogous to C, which we denote by C_0, must be a 3-coboundary. Thus

$$\begin{aligned} C(j_1, j_2, j_3) &= (-1)^{\langle \lambda_{j_1}, \varepsilon(j_2, j_3) \rangle} c(\lambda_{j_1}, \varepsilon(j_2, j_3))^{-1} C_0(j_1, j_2, j_3) \\ &= h(j_2, j_3, j_1)^{-1} C_0(j_1, j_2, j_3) \end{aligned} \tag{12.44}$$

(recall (12.25)) and C is cohomologous to

$$\begin{aligned} h_0 : L/L_0 \times L/L_0 \times L/L_0 &\to \langle \omega_q \rangle \\ (j_1, j_2, j_3) &\mapsto h(j_2, j_3, j_1)^{-1}. \end{aligned} \tag{12.45}$$

Note that the cohomology class of the cocycle h_0 is independent of our choice of c. The cocycle h_0 is not quite the same as h. To complete the proof, it suffices to show that h and h_0 are cohomologous.

For this, we use our assumption (12.40) to extend the biadditive map A given by (12.38) to the (symmetric) biadditive map

$$\begin{aligned} L \times L &\to \langle \omega_q \rangle \\ (\alpha, \beta) &\mapsto e^{i\pi \langle \alpha, \beta \rangle} \in \langle \omega_q \rangle. \end{aligned} \tag{12.46}$$

Consider the biadditive extension of the map $k(\cdot,\cdot)^{-1}$ (recall (12.26)), viewed as a map $L_0 \times L \to \langle\omega_q\rangle$, given by

$$\begin{aligned} K : L \times L &\to \langle\omega_q\rangle \\ (\alpha,\beta) &\mapsto e^{i\pi\langle\alpha,\beta\rangle}c(\beta,\alpha) \in \langle\omega_q\rangle, \end{aligned} \tag{12.47}$$

and note that

$$K(\alpha,\beta)K(\beta,\alpha) = 1 \quad \text{for} \quad \alpha \in L_0,\ \beta \in L. \tag{12.48}$$

We use K to define a 2-cochain

$$\begin{aligned} \gamma : L/L_0 \times L/L_0 &\to \langle\omega_q\rangle \\ (j_1,j_2) &\mapsto K(\lambda_{j_2},\lambda_{j_1}) \end{aligned} \tag{12.49}$$

(cf. (12.22)), and with the help of (12.48) we can compute directly that

$$\begin{aligned} (d\gamma)(j_1,j_2,j_3) &= k(\varepsilon(j_1,j_2),j_3)k(\varepsilon(j_2,j_3),j_1) \\ &= h(j_1,j_2,j_3)^{-1}h_0(j_1,j_2,j_3). \end{aligned} \tag{12.50}$$

Thus, $[h] = [h_0]$, completing the proof. □

Remark 12.16: The fact that the cohomology class h_0 (12.45) is independent of the choice of c implies immediately that if c_0 is any alternating bilinear map $L \times L \to \langle\omega_q\rangle$ which vanishes on $L_0 \times L_0$, then the (well-defined) function

$$\begin{aligned} L/L_0 \times L/L_0 \times L/L_0 &\to \langle\omega_q\rangle \\ (j_1,j_2,j_3) &\mapsto c_0(\varepsilon(j_2,j_3),j_1) \end{aligned} \tag{12.51}$$

is a coboundary.

Remark 12.17: Theorem 12.15 shows that (under the mild assumption (12.40)) the class $[h]$ associated with the Jacobi identity (12.19) depends only on the data entering into the exact sequence (12.37) and the action of L on $\hat{L}_0$, in particular, the pair of lattices $L_0 \subset L$ together with the biadditive map A given by (12.38). On the other hand, given this data, one forms an associated cocycle by combining it appropriately with a cocycle for the extension $L \to L/L_0$ and an alternating biadditive map on $L \times L$ agreeing with $A|_{L_0 \times L_0}$. The cohomology class is of course related to the failure of the map A to extend to an alternating biadditive map on $L \times L$.

Remark 12.18: We have constantly been assuming the existence of an alternating $\mathbf{Z}$-bilinear function c: $L \times L \to \langle \omega_q \rangle$ such that $c(\alpha, \beta) = (-1)^{\langle \alpha, \beta \rangle}$ for $\alpha, \beta \in L_0$. But it is easy to see that by choosing the even integer q large enough, such a function c always exists: Write the given alternating (and symmetric) function on $L_0 \times L_0$ in the form $\delta(\alpha, \beta)\delta(\beta, \alpha)^{-1}$, where δ: $L_0 \times L_0 \to \langle -1 \rangle$ is a $\mathbf{Z}$-bilinear cocycle for $\hat{L}_0$ (cf. [FLM3], Section 5.2). View δ as a homomorphism $L_0 \otimes_{\mathbf{Z}} L_0 \to \langle -1 \rangle$ and extend it into a homomorphism δ_1: $L \otimes_{\mathbf{Z}} L \to \langle \omega_q \rangle$. (Note that $L_0 \otimes_{\mathbf{Z}} L_0 \hookrightarrow L \otimes_{\mathbf{Z}} L$.) Then regard δ_1 as a $\mathbf{Z}$-bilinear function and take

$$c(\alpha, \beta) = \delta_1(\alpha, \beta)\delta_1(\beta, \alpha)^{-1}. \tag{12.52}$$

Remark 12.19: On the other hand, as was pointed out in Remark 12.14, the biadditive map A (12.38) need *not* extend to an alternating biadditive map $L \times L \to \mathbf{C}^{\times}$.

Now we discuss the important duality and monodromy results – the analogues of the results of Chapters 7 and 8 – associated with the Jacobi identity (12.19), and at the same time, for the purpose of axiomatization, we keep track of the properties used. As in Chapter 9, we see again that suitable variants of virtually all the the arguments and results in Chapters 7 and 8 remain valid. First, the obvious analogues of (7.3), of the generalized commutator and associator relations (7.5) and (7.11) and of Propositions 7.6, 7.8, 7.9 and 7.11 and Remark 7.7 and Corollary 7.10 hold here, with the numerical factors $c(\lambda_{j_2}, \lambda_{j_1})$ and $h(j_1, j_2, j_3)$ inserted appropriately into the relevant formulas. For instance, using the notation of Theorem 12.11, the generalized commutativity and generalized associativity assertions can be expressed as follows (the analogues of (7.23) and (7.33)): Acting on $V(j_3)$,

$$\text{“}\mathcal{Y}_{\lambda_{j_1}}(w_1, z_1)\mathcal{Y}_{\lambda_{j_2}}(w_2, z_2)(z_1 - z_2)^{-\langle \lambda_{j_1}, \lambda_{j_2} \rangle} \cdot \\ \cdot z_1^{-\langle \lambda_{j_1}, \lambda_{j_3} \rangle} z_2^{-\langle \lambda_{j_2}, \lambda_{j_3} \rangle} \tag{12.53}$$

agrees with

$$c(\lambda_{j_2}, \lambda_{j_1})\mathcal{Y}_{\lambda_{j_2}}(w_2, z_2)\mathcal{Y}_{\lambda_{j_1}}(w_1, z_1)(z_2 - z_1)^{-\langle \lambda_{j_1}, \lambda_{j_2} \rangle} \cdot \\ \cdot z_1^{-\langle \lambda_{j_1}, \lambda_{j_3} \rangle} z_2^{-\langle \lambda_{j_2}, \lambda_{j_3} \rangle} \tag{12.54}$$

and with

$$h(j_1, j_2, j_3)\mathcal{Y}_{\lambda_{j_1 + j_2}}(\mathcal{Y}_{\lambda_{j_1}}(w_1, z_1 - z_2)w_2, z_2) \cdot \\ \cdot (z_1 - z_2)^{-\langle \lambda_{j_1}, \lambda_{j_2} \rangle} z_1^{-\langle \lambda_{j_1}, \lambda_{j_3} \rangle} z_2^{-\langle \lambda_{j_2}, \lambda_{j_3} \rangle} \tag{12.55}$$

as operator-valued rational functions," and this assertion implies the Jacobi identity (12.19). Moreover, the analogue of Proposition 7.12, or rather, of Proposition 9.17, holds here, since the proof requires only the $\mathbf{Z}$-bilinearity $\bmod \mathbf{Z}$ (and not $\bmod 2\mathbf{Z}$) of the function $(j_1, j_2) \mapsto \langle \lambda_{j_1}, \lambda_{j_2} \rangle$ on $L/L_0 \times L/L_0$ and only the skew-symmetry (and not the $\mathbf{Z}$-bilinearity) of the function $(j_1, j_2) \mapsto c(\lambda_{j_2}, \lambda_{j_1})$, both of which are true. The analogue of (7.40) or (9.42) becomes: For $j_1, ..., j_n \in L/L_0$, $w_r \in V(j_r)$ for each r, and any permutation $(i_1 \cdots i_n)$ of $(1 \cdots n)$,

$$\begin{aligned} &\text{“}\mathcal{Y}_{\lambda_{j_1}}(w_1, z_1) \cdots \mathcal{Y}_{\lambda_{j_n}}(w_n, z_n) \cdot \\ &\cdot \prod_{k<l} (z_k - z_l)^{-\langle \lambda_{j_k}, \lambda_{j_l} \rangle} \prod_{m=1}^{n} z_m^{-\lambda_{jm}} \quad \text{agrees with} \\ (12.56) \qquad \prod_{k<l,\, i_k > i_l} &c(\lambda_{j_{i_k}}, \lambda_{j_{i_l}}) \mathcal{Y}_{\lambda_{j_{i_1}}}(w_{i_1}, z_{i_1}) \cdots \mathcal{Y}_{\lambda_{j_{i_n}}}(w_{i_n}, z_{i_n}) \cdot \\ &\cdot \prod_{k<l} (z_{i_k} - z_{i_l})^{-\langle \lambda_{j_{i_k}}, \lambda_{j_{i_l}} \rangle} \prod_{m=1}^{n} z_m^{-\lambda_{jm}} \end{aligned}$$

as operator-valued rational functions." The analogue of Remark 7.13 is valid.

We note that the "creation property" holds on V_L for the intertwining operator $\mathcal{Y}_{\lambda_j}$: For $j \in L/L_0$ and $w \in V(j)$,

$$(12.57) \qquad \mathcal{Y}_{\lambda_j}(w, z)\mathbf{1} \in V[[z]] \quad \text{and} \quad \lim_{z \to 0} \mathcal{Y}_{\lambda_j}(w, z)\mathbf{1} = w.$$

Using this, we see that the analogues of Remarks 7.14 and 7.15 hold. Moreover,

$$(12.58) \qquad \mathcal{Y}_{\lambda_j}(w, z)\mathbf{1} = e^{zL(-1)}w.$$

Proposition 7.16 – the fact that generalized commutativity implies generalized associativity – also carries over, in a particularly interesting way, and we shall discuss this in some detail: First, the analogue of (7.48) states that for $j_i \in L/L_0$ and $w_i \in V(j_i)$ $(i = 1, 2, 3)$ and $w' \in V'_L$,

$$(12.59) \qquad \begin{aligned} &\langle w', \mathcal{Y}_{\lambda_{j_3}}(w_3, z_3) \mathcal{Y}_{\lambda_{j_1 + j_2}}(\mathcal{Y}_{\lambda_{j_1}}(w_1, z_0) w_2, z_2)\mathbf{1} \rangle \\ &= \langle w', \mathcal{Y}_{\lambda_{j_3}}(w_3, z_3) \mathcal{Y}_{\lambda_{j_1}}(w_1, z_0 + z_2) \mathcal{Y}_{\lambda_{j_2}}(w_2, z_2)\mathbf{1} \rangle, \end{aligned}$$

and multiplication by $z_0^{-\langle \lambda_{j_1}, \lambda_{j_2} \rangle} z_3^{-\langle \lambda_{j_1}, \lambda_{j_3} \rangle} z_3^{-\langle \lambda_{j_2}, \lambda_{j_3} \rangle}$ turns this into a series involving integral powers of z_0, z_2 and z_3. We shall use the following version of (7.49):

$$\begin{aligned}(12.60)\quad & c(\lambda_{j_3},\lambda_{j_1})c(\lambda_{j_3},\lambda_{j_2})\langle w',\mathcal{Y}_{\lambda_{j_3}}(w_3,z_3)\mathcal{Y}_{\lambda_{j_1}}(w_1,z_1)\mathcal{Y}_{\lambda_{j_2}}(w_2,z_2)\mathbf{1}\rangle\cdot\\ & \cdot(z_3-z_1)^{-\langle\lambda_{j_1},\lambda_{j_3}\rangle}(z_3-z_2)^{-\langle\lambda_{j_2},\lambda_{j_3}\rangle}(z_1-z_2)^{-\langle\lambda_{j_1},\lambda_{j_2}\rangle}\\ & =\iota_{312}f(z_1,z_2,z_3)\end{aligned}$$

where f is as in (7.50), and where the permutation-independence of f involves the numerical factors given in (12.56).

Formula (7.51) becomes: For $m\in\mathbf{Z}$,

$$\begin{aligned}(12.61)\quad & \iota_{23}^{-1}\Big(\langle w',\mathcal{Y}_{\lambda_{j_1+j_2}}(\mathcal{Y}_{\lambda_{j_1}}(w_1,z_0)w_2,z_2)\mathcal{Y}_{\lambda_{j_3}}(w_3,z_3)\mathbf{1}\rangle\cdot\\ & \cdot z_0^{-\langle\lambda_{j_1},\lambda_{j_2}\rangle}(z_2-z_3)^{-\langle\lambda_{j_1},\lambda_{j_3}\rangle-\langle\lambda_{j_2},\lambda_{j_3}\rangle+m}\Big)\\ & =(-1)^{\langle\varepsilon(j_1,j_2),\lambda_{j_3}\rangle+m}c(\lambda_{j_3},\lambda_{j_1+j_2})\cdot\\ & \cdot\iota_{32}^{-1}\Big(\langle w',\mathcal{Y}_{\lambda_{j_3}}(w_3,z_3)\mathcal{Y}_{\lambda_{j_1+j_2}}(\mathcal{Y}_{\lambda_{j_1}}(w_1,z_0)w_2,z_2)\mathbf{1}\rangle\cdot\\ & \cdot z_0^{-\langle\lambda_{j_1},\lambda_{j_2}\rangle}(z_3-z_2)^{-\langle\lambda_{j_1},\lambda_{j_3}\rangle-\langle\lambda_{j_2},\lambda_{j_3}\rangle+m}\Big).\end{aligned}$$

The computation (7.52) thus gives

$$\begin{aligned}(12.62)\quad & \iota_{23}^{-1}\Big(\langle w',\mathcal{Y}_{\lambda_{j_1+j_2}}(\mathcal{Y}_{\lambda_{j_1}}(w_1,z_0)w_2,z_2)\mathcal{Y}_{\lambda_{j_3}}(w_3,z_3)\mathbf{1}\rangle\cdot\\ & \cdot z_0^{-\langle\lambda_{j_1},\lambda_{j_2}\rangle}(z_2-z_3)^{-\langle\lambda_{j_2},\lambda_{j_3}\rangle}(z_2-z_3+z_0)^{-\langle\lambda_{j_1},\lambda_{j_3}\rangle}\Big)\\ & =(-1)^{\langle\varepsilon(j_1,j_2),\lambda_{j_3}\rangle}c(\lambda_{j_3},\lambda_{j_1+j_2})\cdot\\ & \cdot\iota_{32}^{-1}\Big(\langle w',\mathcal{Y}_{\lambda_{j_3}}(w_3,z_3)\mathcal{Y}_{\lambda_{j_1+j_2}}(\mathcal{Y}_{\lambda_{j_1}}(w_1,z_0)w_2,z_2)\mathbf{1}\rangle\cdot\\ & \cdot z_0^{-\langle\lambda_{j_1},\lambda_{j_2}\rangle}(z_3-z_2)^{-\langle\lambda_{j_2},\lambda_{j_3}\rangle}(z_3-z_2-z_0)^{-\langle\lambda_{j_1},\lambda_{j_3}\rangle}\Big)\\ & =(-1)^{\langle\varepsilon(j_1,j_2),\lambda_{j_3}\rangle}c(\lambda_{j_3},\varepsilon(j_1,j_2))^{-1}\iota_{32}^{-1}\left(\iota_{312}f(z_1,z_2,z_3)|_{z_1=z_0+z_2}\right)\\ & =h(j_1,j_2,j_3)^{-1}\iota_{32}^{-1}\left(\iota_{312}f(z_1,z_2,z_3)|_{z_1=z_0+z_2}\right)\end{aligned}$$

(see Remark 12.20 below), and as in (7.52)-(7.53), the application of ι_{23} yields

$$(12.63)\qquad h(j_1,j_2,j_3)^{-1}\iota_{230}f(z_0+z_2,z_2,z_3).$$

But as in (7.54), we also have

$$\begin{aligned}(12.64)\quad & \iota_{023}f(z_0+z_2,z_2,z_3)\\ & =\langle w',\mathcal{Y}_{\lambda_{j_1}}(w_1,z_0+z_2)\mathcal{Y}_{\lambda_{j_2}}(w_2,z_2)\mathcal{Y}_{\lambda_{j_3}}(w_3,z_3)\mathbf{1}\rangle\cdot\\ & \cdot(z_0+z_2-z_3)^{-\langle\lambda_{j_1},\lambda_{j_3}\rangle}(z_2-z_3)^{-\langle\lambda_{j_2},\lambda_{j_3}\rangle}z_0^{-\langle\lambda_{j_1},\lambda_{j_2}\rangle},\end{aligned}$$

and setting $z_3 = 0$ as in (7.55)-(7.56) we obtain the appropriate associativity result – the equality of (12.53) and (12.55). Thus we have the analogues of Proposition 7.16 and of Remarks 7.17 and 7.18.

Remark 12.20: Note how the factor $h(j_1, j_2, j_3)$ arises from the product of the failures of $\mathbf{Z}$-bilinearity of the two functions from $L/L_0 \times L/L_0$ given by $\langle \lambda_{j_1}, \lambda_{j_2} \rangle$ (see (12.61)) and $c(\lambda_{j_1}, \lambda_{j_2})$. Compare with Remark 12.12.

Just as at the end of Chapter 9, we see easily that all the results on analytic continuation and one-dimensional braid group representations in Chapter 8 carry over to the present situation, with only the following modifications: The setting of (12.56) is of course used, in place of the setting of Proposition 7.12 or 9.17. The appropriate analogue of the expression (8.11) includes the numerical factor in (12.56), and the left-hand side of the analogue of (8.13) and the right-hand sides of the analogues of (8.21) and (8.22) are multiplied by $c(\lambda_{j_{i+1}}, \lambda_{j_i})$. Because of the specialization (8.24), which now becomes the specialization

$$j_1 = \cdots = j_n = j \in L/L_0, \tag{12.65}$$

and because of the alternating property $c(\lambda_j, \lambda_j) = 1$, the conclusions on braid group representations carry over, and we find as in Proposition 8.3 that on the analogue of the one-dimensional space given by (8.29), the braid group B_n acts by:

$$b_k \mapsto e^{i\pi\langle \lambda_j, \lambda_j \rangle} \quad \text{for} \quad k = 1, ..., n-1. \tag{12.66}$$

As in Remark 8.4, the Hecke algebra $H_n(e^{i\pi\langle \lambda_j, \lambda_j \rangle})$ acts correspondingly.

We shall now prepare to axiomatize certain important features of our situation. Let us denote our abelian group L/L_0 by G:

$$G = L/L_0. \tag{12.67}$$

Denoting elements of G by $g_1, g_2, ...,$ we consider the following three functions on $G \times G$:

$$\begin{aligned} p : G \times G &\rightarrow \left(\tfrac{1}{T}\mathbf{Z}\right)/2\mathbf{Z} \\ (g_1, g_2) &\mapsto -\langle \lambda_{g_1}, \lambda_{g_2} \rangle + 2\mathbf{Z} \end{aligned} \tag{12.68}$$

$$\begin{aligned} r : G \times G &\rightarrow \langle \omega_q \rangle \\ (g_1, g_2) &\mapsto c(\lambda_{g_1}, \lambda_{g_2})^{-1} \end{aligned} \tag{12.69}$$

$$\begin{aligned} s : G \times G &\to \mathbf{C}^{\times} \\ (g_1, g_2) &\mapsto r(g_1, g_2) e^{i\pi p(g_1, g_2)}, \end{aligned} \tag{12.70}$$

where T is a positive integer such that

$$\langle L, L \rangle \subset \frac{1}{T}\mathbf{Z}; \tag{12.71}$$

we do not assume any relation between T and our positive integer q, such as would be implied by (12.40). Note that p is symmetric, i.e.,

$$p(g_1, g_2) = p(g_2, g_1), \tag{12.72}$$

and biadditive mod $\mathbf{Z}$, i.e.,

$$p(g_1 + g_2, g_3) - p(g_1, g_3) - p(g_2, g_3) \in \mathbf{Z}/2\mathbf{Z} \tag{12.73}$$

and that r is skew-symmetric, in the sense that

$$r(g_1, g_2) r(g_2, g_1) = 1, \tag{12.74}$$

and alternating, in the sense that

$$r(g_1, g_1) = 1. \tag{12.75}$$

Also,

$$p(g_1, 0) = 0, \tag{12.76}$$

$$r(g_1, 0) = 1. \tag{12.77}$$

Consider the function h (recall (12.20)), which is given in terms of our data by

$$\begin{aligned} h : G \times G \times G &\to \langle \omega_q \rangle \\ (g_1, g_2, g_3) &\mapsto s(g_1 + g_2, g_3)^{-1} s(g_1, g_3) s(g_2, g_3) \\ &\quad = r(g_1 + g_2, g_3)^{-1} r(g_1, g_3) r(g_2, g_3) \cdot \\ &\qquad \cdot (-1)^{p(g_1+g_2, g_3) - p(g_1, g_3) - p(g_2, g_3)}. \end{aligned} \tag{12.78}$$

Note that h in fact has values in $\langle \omega_q \rangle$ (by the biadditivity of p mod $\mathbf{Z}$ (12.73)), that h is symmetric in the first two variables, i.e.,

$$h(g_1, g_2, g_3) = h(g_2, g_1, g_3), \tag{12.79}$$

and that h is a 2-cocycle on $G \times G$ in the first two variables with the third variable held fixed, i.e.,

$$(12.80)\qquad \begin{aligned} &h(g_1, g_2, g_4)h(g_1, g_2 + g_3, g_4)^{-1}\cdot \\ &\cdot h(g_1 + g_2, g_3, g_4)h(g_2, g_3, g_4)^{-1} = 1. \end{aligned}$$

Thus the condition that h be a 3-cocycle on G with values in $\langle \omega_q \rangle$, i.e., that

$$(12.81)\qquad \begin{aligned} &h(g_1, g_2, g_3)h(g_1, g_2, g_3 + g_4)^{-1}h(g_1, g_2 + g_3, g_4)\cdot \\ &\cdot h(g_1 + g_2, g_3, g_4)^{-1}h(g_2, g_3, g_4) = 1, \end{aligned}$$

is equivalent to the additivity of h in the third variable, i.e.,

$$(12.82)\qquad h(g_1, g_2, g_3 + g_4) = h(g_1, g_2, g_3)h(g_1, g_2, g_4),$$

which is true in our situation. Also, while the function s is not necessarily skew-symmetric since $p(\cdot,\cdot)$ is not necessarily $\mathbf{Z}/2\mathbf{Z}$-valued (T might be greater than 1), we still have

$$(12.83)\qquad h(g_1, g_2, g_3) = s(g_3, g_1 + g_2)s(g_3, g_1)^{-1}s(g_3, g_2)^{-1},$$

by (12.72), (12.73) and (12.74), as if s *were* skew-symmetric.

Moreover,

$$(12.84)\qquad \begin{aligned} &s(g_1 + g_2, g_1 + g_2)s(g_1, g_1)^{-1}s(g_2, g_2)^{-1} \\ &= e^{i\pi(p(g_1+g_2, g_1+g_2) - p(g_1, g_1) - p(g_2, g_2))} = e^{2\pi i p(g_1, g_2)} \end{aligned}$$

(even though p is not in general biadditive mod $2\mathbf{Z}$), which follows either from (12.68) and (12.69), or alternatively, from the following abstract argument using (12.78), (12.82) and (12.83), together with (12.72), (12.74) and (12.75):

$$(12.85)\qquad \begin{aligned} s(g_1 + g_2, g_1 + g_2) &= s(g_1 + g_2, g_1)s(g_1 + g_2, g_2)h(g_1, g_2, g_1 + g_2) \\ &= s(g_1, g_1)s(g_2, g_1)s(g_1, g_2)s(g_2, g_2) \\ &= s(g_1, g_1)s(g_2, g_2)e^{2\pi i p(g_1, g_2)}. \end{aligned}$$

By (12.73), the expression (12.84) is biadditive as a function of g_1 and g_2. In addition, the calculation of $h(g_1, -g_1, g_1)h(g_1, -g_1, -g_1)$ (which equals 1 by (12.82)), by means of (12.78) for one factor and (12.83) for the other, shows that

$$(12.86)\qquad s(g_1, g_1) = s(-g_1, -g_1).$$

Using the biadditivity (12.84), we see inductively that

$$s(ng_1, ng_1) = s(g_1, g_1)^{n^2} \quad \text{for} \quad n \in \mathbf{Z}, \tag{12.87}$$

which together with the biadditivity tells us that the function

$$\begin{array}{rcl} G & \to & \mathbf{C}^{\times} \\ g & \mapsto & s(g,g) = e^{i\pi p(g,g)} \end{array} \tag{12.88}$$

is a quadratic form with associated bilinear form given by $e^{2\pi i p(g_1,g_2)}$.

Remark 12.21: If we may choose $T = \frac{1}{2}q$, as in the assumption (12.40) in Theorem 12.15, then h is actually a 2-coboundary on $G \times G$ in the first two variables with values in $\langle \omega_q \rangle$, not just a cocycle, since in this case,

$$s : G \times G \to \langle \omega_q \rangle. \tag{12.89}$$

Before we formalize this structure axiomatically, note that it is reasonable to allow the following renormalization of our intertwining operators: In the setting of Theorem 12.11, let us multiply each $\mathcal{Y}_{\lambda_{g_1}}|_{V(g_2)}$ $(g_1, g_2 \in G = L/L_0)$ by

$$f(g_1, g_2) \in \mathbf{C}^{\times} \tag{12.90}$$

(cf. Proposition 12.3), and set

$$\mathcal{X}_{g_1,g_2}(w,z) = f(g_1,g_2)\mathcal{Y}_{\lambda_{g_1}}(w,z)|_{V(g_2)} \text{ for } w \in V(g_1). \tag{12.91}$$

Assume that

$$f(0,g) = 1 \quad \text{for} \quad g \in G, \tag{12.92}$$

corresponding to keeping the module action unmodified, i.e.,

$$\mathcal{X}_{0,g}(w,z) = \mathcal{Y}_{\lambda_0}(w,z)|_{V(g)} = Y(w,z)|_{V(g)} \tag{12.93}$$

for $w \in V(0)$, and that

$$f(g,0) = 1 \quad \text{for} \quad g \in G, \tag{12.94}$$

corresponding to taking that intertwining operator of type $\begin{bmatrix} & 0 & \\ g & & g \end{bmatrix}$ which satisfies the "creation property" (9.12) (see (12.3)), i.e.,

$$\lim_{z \to 0} \mathcal{X}_{g,0}(w,z)\iota(1) = \lim_{z \to 0} Y(w,z)\iota(1) = w \quad \text{for} \quad w \in V(g). \tag{12.95}$$

Then the Jacobi identity (12.19), written using the operators $\mathcal{X}_{g_1,g_2}$ in place of $\mathcal{Y}_{\lambda_{g_1}}|_{V(g_2)}$, asserts: For $g_1, g_2, g_3 \in G = L/L_0$,

$$z_0^{-1}\left(\frac{z_1-z_2}{z_0}\right)^{p(g_1,g_2)}\delta\left(\frac{z_1-z_2}{z_0}\right)\mathcal{X}_{g_1,g_2+g_3}(w_1,z_1)\mathcal{X}_{g_2,g_3}(w_2,z_2)$$

$$-f(g_2,g_3)f(g_2,g_1+g_3)^{-1}f(g_1,g_2+g_3)f(g_1,g_3)^{-1}s(g_1,g_2)\cdot$$

$$\cdot z_0^{-1}\left(\frac{z_2-z_1}{e^{i\pi}z_0}\right)^{p(g_1,g_2)}\delta\left(\frac{z_2-z_1}{-z_0}\right)\mathcal{X}_{g_2,g_1+g_3}(w_2,z_2)\mathcal{X}_{g_1,g_3}(w_1,z_1) \tag{12.96}$$

$$= f(g_2,g_3)f(g_1+g_2,g_3)^{-1}f(g_1,g_2+g_3)f(g_1,g_2)^{-1}h(g_1,g_2,g_3)\cdot$$

$$\cdot z_2^{-1}\delta\left(\frac{z_1-z_0}{z_2}\right)\mathcal{X}_{g_1+g_2,g_3}(\mathcal{X}_{g_1,g_2}(w_1,z_0)w_2,z_2)\left(\frac{z_1-z_0}{z_2}\right)^{-p(g_1,g_3)}$$

on $V(g_3)$ (recall Remark 12.12 and (12.68)-(12.70)). Thus the right-hand side has gained the factor

$$(df)(g_1,g_2,g_3) = f(g_2,g_3)f(g_1+g_2,g_3)^{-1}\cdot \\ \cdot f(g_1,g_2+g_3)f(g_1,g_2)^{-1}, \tag{12.97}$$

the coboundary of $f(\cdot,\cdot)$, and the second term on left-hand side has gained the factor

$$f(g_2,g_3)f(g_2,g_1+g_3)^{-1}f(g_1,g_2+g_3)f(g_1,g_3)^{-1}. \tag{12.98}$$

Set

$$F(g_1,g_2,g_3) = (df)(g_1,g_2,g_3)h(g_1,g_2,g_3), \tag{12.99}$$

$$\Omega(g_1,g_2) = f(g_1,g_2)f(g_2,g_1)^{-1}s(g_1,g_2) \tag{12.100}$$

and

$$B(g_1,g_2,g_3) = f(g_2,g_3)f(g_2,g_1+g_3)^{-1}f(g_1,g_2+g_3)\cdot \\ \cdot f(g_1,g_3)^{-1}s(g_1,g_2) \tag{12.101}$$

for $g_1, g_2, g_3 \in G$ (using notation consistent with that of [MS]). Then B can be expressed in terms of F and Ω by

$$B(g_1,g_2,g_3) = F(g_2,g_1,g_3)^{-1}\Omega(g_1,g_2)F(g_1,g_2,g_3), \tag{12.102}$$

by (12.79). Observe that the quadratic form $s(g,g)$ (12.88), which we now denote by $q(g)$, can be expressed in terms of Ω by

$$q(g) = \Omega(g,g), \tag{12.103}$$

and that its associated bilinear form corresponds to the function

$$\begin{aligned} b : G \times G &\to \left(\tfrac{1}{T}\mathbf{Z}\right)/\mathbf{Z} \\ (g_1, g_2) &\mapsto p(g_1, g_2) \bmod \mathbf{Z}, \end{aligned} \tag{12.104}$$

since

$$q(g_1 + g_2)q(g_1)^{-1}q(g_2)^{-1} = e^{2\pi i b(g_1,g_2)}. \tag{12.105}$$

(Note that $p(\cdot,\cdot)$ is $(\frac{1}{T}\mathbf{Z})/2\mathbf{Z}$-valued while $b(\cdot,\cdot)$ is $(\frac{1}{T}\mathbf{Z})/\mathbf{Z}$-valued.) Thus the Jacobi identity (12.96) can be rewritten as

$$\begin{aligned} & z_0^{-1}\left(\frac{z_1 - z_2}{z_0}\right)^{b(g_1,g_2)}\delta\left(\frac{z_1 - z_2}{z_0}\right)\mathcal{X}_{g_1,g_2+g_3}(w_1, z_1)\mathcal{X}_{g_2,g_3}(w_2, z_2) \\ & \quad - B(g_1, g_2, g_3)z_0^{-1}\left(\frac{z_2 - z_1}{e^{i\pi}z_0}\right)^{b(g_1,g_2)}\delta\left(\frac{z_2 - z_1}{-z_0}\right)\cdot \\ & \quad \cdot \mathcal{X}_{g_2,g_1+g_3}(w_2, z_2)\mathcal{X}_{g_1,g_3}(w_1, z_1) \\ & = F(g_1, g_2, g_3)z_2^{-1}\delta\left(\frac{z_1 - z_0}{z_2}\right)\mathcal{X}_{g_1+g_2,g_3}(\mathcal{X}_{g_1,g_2}(w_1, z_0)w_2, z_2)\cdot \\ & \quad \cdot \left(\frac{z_1 - z_0}{z_2}\right)^{-b(g_1,g_3)}. \end{aligned} \tag{12.106}$$

Note that F is a normalized 3-cocycle on G with values in $\mathbf{C}^\times$ (see (12.81)):

$$F(g_1, g_2, 0) = F(g_1, 0, g_3) = F(0, g_2, g_3) = 1 \tag{12.107}$$

$$\begin{aligned} & F(g_1, g_2, g_3)F(g_1, g_2, g_3 + g_4)^{-1}F(g_1, g_2 + g_3, g_4)\cdot \\ & \quad \cdot F(g_1 + g_2, g_3, g_4)^{-1}F(g_2, g_3, g_4) = 1. \end{aligned} \tag{12.108}$$

Moreover, Ω is normalized in the sense that

$$\Omega(g_1, 0) = \Omega(0, g_2) = 1, \tag{12.109}$$

by (12.72), (12.74), (12.76), (12.77), (12.92) and (12.94), and F and Ω satisfy the conditions

$$\begin{aligned} & F(g_1, g_2, g_3)^{-1}\Omega(g_1, g_2 + g_3)F(g_2, g_3, g_1)^{-1} \\ & \quad = \Omega(g_1, g_2)F(g_2, g_1, g_3)^{-1}\Omega(g_1, g_3), \end{aligned} \tag{12.110}$$

$$F(g_1, g_2, g_3)\Omega(g_1 + g_2, g_3)F(g_3, g_1, g_2) = \Omega(g_2, g_3)F(g_1, g_3, g_2)\Omega(g_1, g_3). \tag{12.111}$$

In fact, the pair (h, s) in place of (F, Ω) satisfies these conditions by (12.78) and (12.83) (the two expressions of h as failures of bilinearity of s) together with the symmetry condition (12.79), and it is easy to check directly that the pair consisting of df and the function, say $a_f(\cdot, \cdot)$, given by $(g_1, g_2) \mapsto f(g_1, g_2)f(g_2, g_1)^{-1}$ satisfies the two conditions (even without the use of the normalization conditions (12.92) and (12.94)). The relations (12.107)-(12.111) assert that the pair (F, Ω) is a *(normalized) abelian 3-cocycle* for G with coefficients in $\mathbf{C}^\times$, and for a function $f : G \times G \to \mathbf{C}^\times$, the pair $((df)(\cdot, \cdot, \cdot), a_f(\cdot, \cdot))$ is the *(abelian) coboundary* of f; if f satisfies (12.92) and (12.94) then df satisfies (12.107) and a_f satisfies (12.109) (see [ML]). The normalization condition (12.109) follows from (12.107), (12.110) and (12.111). Any 3-cocycle is cohomologous to a normalized one.

Remark 12.22: The cohomology theory under consideration here is the "cohomology of abelian groups" introduced by Eilenberg-Mac Lane ([E], [ML], [EML1], [EML2]; see also [JS]). This connection was pointed out to us by M. Tierney. The *trace* of an abelian 3-cocycle (F, Ω) is the function given by (12.103), which in general is a quadratic form, as is staightfoward to verify. The associated bilinear form is: $(g_1, g_2) \mapsto \Omega(g_1, g_2)\Omega(g_2, g_1)$. It is also known that trace induces an isomorphism between the abelian third cohomology group and the group of quadratic forms from (in our case) G to $\mathbf{C}^\times$ (see [ML], [JS]). The 3-cocycle relation (12.108) can be interpreted as a pentagon identity, and the abelian 3-cocycle relations (12.110)-(12.111) can be interpreted as hexagon identities (cf. [JS]). The material of the present chapter is also intimately related to Appendix E of [MS], in which these pentagon and hexagon relations are considered for a special class of conformal field theories – those in which the fusion rules are based on abelian groups. In the present chapter, we are essentially considering the construction of vertex-operator-algebraic structures which implement these relations. In [MS], pentagon and hexagon relations, together with other relations, are studied in an abstract way for very general conformal field theories.

We are now ready to formulate the following notion of "abelian intertwining algebra," which clearly generalizes that of generalized vertex algebra (recall Chapter 9):

Let G be an abelian group equipped with two functions,

$$F: G \times G \times G \to \mathbf{C}^{\times}, \tag{12.112}$$

$$\Omega: G \times G \to \mathbf{C}^{\times} \tag{12.113}$$

such that (12.107)-(12.111) hold, i.e., (F, Ω) is a normalized abelian 3-cocycle for G with coefficients in $\mathbf{C}^{\times}$. Denote by

$$\begin{aligned} q: G &\to \mathbf{C}^{\times} \\ g &\mapsto q(g) = \Omega(g, g) \end{aligned} \tag{12.114}$$

the corresponding quadratic form, and by $e^{2\pi i b(\cdot,\cdot)}$ its associated (symmetric) bilinear form, where we have

$$b: G \times G \to \mathbf{C}/\mathbf{Z} \tag{12.115}$$

and

$$q(g_1+g_2)q(g_1)^{-1}q(g_2)^{-1} = e^{2\pi i b(g_1,g_2)} \;\; (= \Omega(g_1, g_2)\Omega(g_2, g_1)) \tag{12.116}$$

for $g_1, g_2 \in G$ (cf. (12.85), (12.88)). Define

$$B: G \times G \times G \to \mathbf{C}^{\times} \tag{12.117}$$

by (12.102). Let T be a positive integer. Assume that b is restricted to the following values:

$$b: G \times G \to \left(\frac{1}{T}\mathbf{Z}\right)/\mathbf{Z}. \tag{12.118}$$

(We could also choose to assume that the values of F and Ω lie in a finite (cyclic) subgroup of $\mathbf{C}^{\times}$.) An *abelian intertwining algebra* of *level* T associated with G, F and Ω is a vector space V with a $\frac{1}{T}\mathbf{Z}$-gradation and a G-gradation, denoted by (9.6), compatible in the sense of (9.7), equipped with a linear map

$$\begin{aligned} V &\to (\operatorname{End} V)[[z^{\frac{1}{T}}, z^{-\frac{1}{T}}]] \\ v &\mapsto Y(v, z) = \sum_{n \in \frac{1}{T}\mathbf{Z}} v_n z^{-n-1} \quad (v_n \in \operatorname{End} V), \end{aligned} \tag{12.119}$$

as in (9.8), and with two distinguished vectors $\mathbf{1} \in V_0^0$, $\omega \in V_2^0$, satisfying the conditions (9.9)-(9.12),

$$Y(v,z)|_{V^h} = \sum_{n \equiv b(g,h) \bmod \mathbf{Z}} v_n z^{-n-1} \quad \text{if} \quad v \in V^g \tag{12.120}$$

(cf. (9.13)), the generalized Jacobi identity

$$z_0^{-1}\left(\frac{z_1-z_2}{z_0}\right)^{b(g_1,g_2)} \delta\left(\frac{z_1-z_2}{z_0}\right) Y(v_1,z_1)Y(v_2,z_2)v_3$$

$$-B(g_1,g_2,g_3)z_0^{-1}\left(\frac{z_2-z_1}{e^{i\pi}z_0}\right)^{b(g_1,g_2)} \delta\left(\frac{z_2-z_1}{-z_0}\right) Y(v_2,z_2)Y(v_1,z_1)v_3 \tag{12.121}$$

$$= F(g_1,g_2,g_3)z_2^{-1}\delta\left(\frac{z_1-z_0}{z_2}\right) Y(Y(v_1,z_0)v_2,z_2)\left(\frac{z_1-z_0}{z_2}\right)^{-b(g_1,g_3)} v_3$$

for $g_1,g_2,g_3 \in G$ and $v_i \in V^{g_i}$, $i = 1,2,3$, and the conditions (9.16)-(9.20). This structure may be denoted

$$(V, Y, \mathbf{1}, \omega, T, G, F(\cdot,\cdot,\cdot), \Omega(\cdot,\cdot)) \tag{12.122}$$

or, briefly, V, and the expressions $Y(v,z)$ are called *(generalized) vertex operators*. It is possible to have the degenerate case $\omega = 0$, or $V = 0$ or $V = \mathbf{C}\mathbf{1}$.

Remark 12.23: Let $f : G \times G \to \mathbf{C}^\times$ be any function satisfying (12.92) and (12.94). If (as in (12.91)) we renormalize each $Y(w,z)|_{V^{g_2}}$ for $w \in V^{g_1}$ by multiplying it by $f(g_1,g_2)$, we again obtain an abelian intertwining algebra for which the pair (F,Ω) is modified by an abelian coboundary as in (12.99)-(12.100), while the quadratic form q and the bilinear form b remain unchanged.

The two main points now are that on the one hand, we have produced an abelian intertwining algebra, and that on the other hand, our main conclusions concerning duality and braid group representations hold for any such structure. First, we summarize many of the results of this chapter, including Theorem 12.11, as follows:

Theorem 12.24 *As in Remark 9.10 and the original setting of this chapter, let L_0 be an even sublattice of a (rational nondegenerate) lattice L (of the same rank) such that*

$$\langle L, L_0 \rangle \subset \mathbf{Z}, \tag{12.123}$$

let c_0 be an alternating $\mathbf{Z}$-bilinear form from $L \times L$ to $\mathbf{Z}/q\mathbf{Z}$ (q a positive even integer) and use c_0 to define c: $L \times L \to \langle \omega_q \rangle$ as in (3.26), and assume that

$$c(\alpha, \beta) = (-1)^{\langle \alpha, \beta \rangle} \quad \text{for} \quad \alpha, \beta \in L_0. \tag{12.124}$$

Set $G = L/L_0$, let T be a positive integer such that

$$\langle \alpha, \alpha \rangle \in \frac{2}{T}\mathbf{Z} \quad \text{for} \quad \alpha \in L, \tag{12.125}$$

and define p, r, s and h as in (12.68), (12.69), (12.70) and (12.78), with λ_g, $g \in G$, a family of coset representatives for G in L such that $\lambda_0 = 0$. Let f be any function as in (12.90) such that (12.92) and (12.94) hold (for example, $f \equiv 1$) and define F, Ω and B as in (12.99), (12.100) and (12.102). Take $V = V_L$, and consider the usual $\frac{1}{T}\mathbf{Z}$-gradation $V = \coprod V_n$ of V and the G-gradation $V = \coprod V^g$ of V given by (5.56), that is,

$$V^g = V(g) \quad \text{for} \quad g \in G. \tag{12.126}$$

Define the operator Y: $V \to (\operatorname{End} V)[[z^{\frac{1}{T}}, z^{-\frac{1}{T}}]]$ by the condition

$$Y(w, z)|_{V^h} = f(g, h)\mathcal{Y}_{\lambda_g}(w, z)|_{V^h} \text{ for } g, h \in G, \ w \in V^g, \tag{12.127}$$

using the notation (12.3), and take $\mathbf{1}$ and ω as usual. Then $(V, Y, \mathbf{1}, \omega, T, G, F, \Omega)$ is an abelian intertwining algebra of rank equal to rank L. *Moreover, the abelian cohomology class of (F, Ω) is independent of the choice of normalized section $g \mapsto \lambda_g$, and in fact if we replace λ_g by $\lambda_g + \alpha_g$ with $\alpha_g \in L_0$ and $\alpha_0 = 0$, then (F, Ω) is modified as in (12.99)-(12.100) with the function f in (12.99)-(12.100) taken to be $(g_1, g_2) \mapsto c(\lambda_{g_2}, \alpha_{g_1})(-1)^{\langle \alpha_{g_1}, \lambda_{g_2} \rangle}$ (cf. Propositions 12.3 and 12.13).*
□

Next we summarize much of the remaining discussion in this chapter with the following comments and results (cf. Chapter 9):

Remark 12.25: In discussing the notion of abelian intertwining algebra, it is sometimes convenient to lift the function b to a function $\hat{b}$: $G \times G \to \frac{1}{T}\mathbf{Z}$ which is symmetric and normalized by the condition $\hat{b}(g, 0) = 0$ for $g \in G$; cf. Remark 6.1. In the definition of abelian intertwining algebra, b may be replaced by $\hat{b}$.

Remark 12.26: The properties (6.20)-(6.28) and Remark 6.3 hold here.

Remark 12.27: Remark 6.2 is replaced by the corresponding statement but with $m \in \mathbf{Z}$ rather than $m \in 2\mathbf{Z}$.

Remark 12.28: In the definition of abelian intertwining algebra, the following restrictions give precisely the definition of generalized vertex algebra (Chapter 9): (1) F is trivial (so that $\Omega(g_1, g_2) = B(g_1, g_2, g_3)$ is $\mathbf{Z}$-bilinear), and (2) the quadratic form $q(\cdot)$ is given by the diagonal values of some symmetric bilinear form, which must necessarily be a symmetric bilinear square root of the symmetric bilinear form $e^{2\pi i b(\cdot,\cdot)}$ associated to q. The square root form may be written as $e^{i\pi \hat{b}(\cdot,\cdot)}$, where $\hat{b}$ is a lift of b, as in Remark 12.25, and is symmetric and bilinear modulo $2\mathbf{Z}$; we have

$$q(g) = e^{i\pi \hat{b}(g,g)} \quad \text{for} \quad g \in G. \tag{12.128}$$

The symmetric bilinear form of formula (9.1) is given by

$$(g_1, g_2) = \hat{b}(g_1, g_2) \bmod 2\mathbf{Z} \tag{12.129}$$

and the alternating bilinear form of (9.2)-(9.5) is given by

$$c(g_1, g_2) = \Omega(g_1, g_2)e^{-i\pi \hat{b}(g_1,g_2)}; \tag{12.130}$$

the alternating property is the expression of $q(\cdot)$ as the diagonal of $e^{i\pi \hat{b}(\cdot,\cdot)}$. Note that even in this special case of abelian intertwining algebra, corresponding to a generalized vertex algebra, the quadratic form (12.128) may be nontrivial, producing a nontrivial abelian third cohomology class and nontrivial braid group representations (see Theorem 12.37 below); this should be contrasted with the fact that the *ordinary* third cohomology class associated with the (trivial) 3-cocycle F is zero. This same comment still applies in the further special case in which the alternating form $c(\cdot, \cdot)$ is taken to be trivial, as occurs for generalized vertex operator algebras (Chapter 6; recall Remark 9.2); by (12.130), this is precisely the case in which Ω is symmetric and agrees with $e^{i\pi \hat{b}}$:

$$\Omega(g_1, g_2) = e^{i\pi \hat{b}(g_1,g_2)}. \tag{12.131}$$

Remark 12.29: The analogue of Remark 6.5 (or Remark 9.5) holds here: If V^0 satisfies the condition (6.30), then $(V^0, Y, \mathbf{1}, \omega)$ is a vertex algebra. The axiom (12.120) follows from the other axioms together with the following natural analogue of the condition (6.31):

$$V^g = \coprod_{e^{2\pi i n} = q(g)^{-1}} V_n^g \quad \text{for} \quad g \in G; \tag{12.132}$$

this is the condition that reduced modulo $\mathbf{Z}$, the weights associated with V^g should be constrained by $q(g)^{-1}$. Formulas (6.30) and (12.132) hold for the abelian intertwining algebra of Theorem 12.24.

Remark 12.30: We have chosen the term "abelian intertwining algebra" because the algebra is made up of intertwining operators for the vertex algebra V^0 (assuming (6.30) for convenience) acting on the modules V^g ($g \in G$) and because of the one-dimensional braid group representations which result and which play such a central role in the structure (see below).

Remark 12.31: The axioms for abelian intertwining algebra are designed so that appropriate analogues of all the comments earlier in this chapter about the validity of the duality and monodromy results of Chapters 7 and 8 for the structure described in Theorem 12.24 hold in general. Specifically, we have the analogues of (7.3), the generalized commutator and associator relations (7.5) and (7.11), Propositions 7.6, 7.8, 7.9 and 7.11, Remark 7.7 and Corollary 7.10, with the numerical factors $e^{-i\pi\hat{b}(g_1,g_2)}B(g_1,g_2,g_3)$ and $F(g_1,g_2,g_3)$ inserted appropriately. The generalized commutativity and associativity relations can be expressed as follows (cf. (7.23), (7.33) and (12.53)-(12.55)): Let $g_1, g_2, g_3 \in G$ and let $v_1 \in V^{g_1}$, $v_2 \in V^{g_2}$. Acting on V^{g_3},

$$\text{(12.133)} \qquad \text{“}Y(v_1,z_1)Y(v_2,z_2)(z_1-z_2)^{\hat{b}(g_1,g_2)}z_1^{\hat{b}(g_1,g_3)}z_2^{\hat{b}(g_2,g_3)}$$

agrees with

$$\text{(12.134)} \qquad \begin{aligned} &B(g_1,g_2,g_3)Y(v_2,z_2)Y(v_1,z_1)(e^{-i\pi}(z_2-z_1))^{\hat{b}(g_1,g_2)}\cdot \\ &\cdot z_1^{\hat{b}(g_1,g_3)}z_2^{\hat{b}(g_2,g_3)} \end{aligned}$$

and with

$$\text{(12.135)} \qquad \begin{aligned} &F(g_1,g_2,g_3)Y(Y(v_1,z_1-z_2)v_2,z_2)(z_1-z_2)^{\hat{b}(g_1,g_2)}\cdot \\ &\cdot z_1^{\hat{b}(g_1,g_3)}z_2^{\hat{b}(g_2,g_3)} \end{aligned}$$

as operator-valued rational functions," and this assertion implies the Jacobi identity (12.121). (Note that if we apply the generalized commutativity relation again to (12.134), we indeed return to (12.133) because of the relation between b and Ω; recall (12.116).) That is, we have:

Theorem 12.32 *In the definition of abelian intertwining algebra, the Jacobi identity can be replaced by the generalized commutativity and associativity properties. Moreover (as in Proposition 7.11), the vector $v' \in V'$ implicit in the generalized commutativity and associativity assertions may be restricted to range through a fixed dense subset of V'.* □

Next we have:

Theorem 12.33 *Under the hypothesis (9.39), which holds in the situation of Theorem 12.24 or if the $\frac{1}{T}\mathbf{Z}$-grading of V is bounded below, we have the following generalized rationality and commutativity for several vertex operators: Let $g_1, ..., g_{n+1} \in G$ and $v_j \in V^{g_j}$ for $j = 1, ..., n$, and let $i = 1, ..., n-1$. Then acting on $V^{g_{n+1}}$,*

$$\text{“}Y(v_1, z_1) \cdots Y(v_n, z_n) \prod_{k<l} (z_k - z_l)^{\hat{b}(g_k, g_l)} \prod_{m=1}^{n} z_m^{\hat{b}(g_m, g_{n+1})} \text{ agrees with}$$

$$\begin{aligned} &B(g_i, g_{i+1}, g_{i+2} + \cdots + g_{n+1}) Y(v_1, z_1) \cdots Y(v_{i-1}, z_{i-1}) \cdot \\ &\cdot Y(v_{i+1}, z_{i+1}) Y(v_i, z_i) Y(v_{i+2}, z_{i+2}) \cdots Y(v_n, z_n) \cdot \\ &\cdot \prod_{k<l,\,(k,l)\neq(i,i+1)} (z_k - z_l)^{\hat{b}(g_k, g_l)} (e^{-i\pi}(z_{i+1} - z_i))^{\hat{b}(g_i, g_{i+1})} \prod_{m=1}^{n} z_m^{\hat{b}(g_m, g_{n+1})} \end{aligned} \tag{12.136}$$

as operator-valued rational functions.” □

Remark 12.34: We also have the analogues of Remarks 7.13 and 7.14 and of Corollary 7.15.

In the present generality, the argument which derives generalized associativity from generalized commutativity uses the relation (12.111) (cf. Remark 12.20)):

Theorem 12.35 *The analogues of Proposition 7.16 and of Remark 7.17 hold. In particular, in the axioms, the Jacobi identity (12.121) may be replaced by generalized rationality and commutativity – the agreement of (12.133) and (12.134) – and (6.20) and (6.21).*

Proof First recall that (6.26)-(6.28) hold. Let $g_i \in G$ and $v_i \in V^{g_i}$ for $i = 1, 2, 3$, and let $v' \in V'$. Then

$$\begin{aligned} &\langle v', Y(v_3, z_3) Y(Y(v_1, z_0) v_2, z_2) \mathbf{1} \rangle \\ &= \langle v', Y(v_3, z_3) Y(v_1, z_0 + z_2) Y(v_2, z_2) \mathbf{1} \rangle \end{aligned} \tag{12.137}$$

and multiplying either side of (12.137) by $z_0^{\hat{b}(g_1,g_2)} z_3^{\hat{b}(g_1,g_3)} z_3^{\hat{b}(g_2,g_3)}$ turns it into a series involving only integral powers of z_0, z_2 and z_3. By assumption, we may write

$$\begin{aligned}(12.138)\qquad &\langle v', Y(v_1,z_1)Y(v_2,z_2)Y(v_3,z_3)\mathbf{1}\rangle \textstyle\prod_{j<k}(z_j-z_k)^{\hat{b}(g_j,g_k)}\\ &= \iota_{123} f(z_1,z_2,z_3),\end{aligned}$$

where f is as in (7.50), and where the permutation-independence property of f involves certain numerical factors. For $m \in \mathbf{Z}$,

$$\begin{aligned}&\iota_{23}^{-1}(\langle v', Y(Y(v_1,z_0)v_2,z_2)Y(v_3,z_3)\mathbf{1}\rangle\cdot\\ &\quad\cdot z_0^{\hat{b}(g_1,g_2)}(z_2-z_3)^{\hat{b}(g_1+g_2,g_3)+m})\\ (12.139)\qquad &= \Omega(g_1+g_2,g_3)\iota_{32}^{-1}(\langle v', Y(v_3,z_3)Y(Y(v_1,z_0)v_2,z_2)\mathbf{1}\rangle\cdot\\ &\quad\cdot z_0^{\hat{b}(g_1,g_2)}(e^{-i\pi}(z_3-z_2))^{\hat{b}(g_1+g_2,g_3)+m}).\end{aligned}$$

Thus, from the bilinearity of $\hat{b}$ modulo $\mathbf{Z}$, we have, just as in (7.52),

$$\begin{aligned}&\iota_{23}^{-1}(\langle v', Y(Y(v_1,z_0)v_2,z_2)Y(v_3,z_3)\mathbf{1}\rangle z_0^{\hat{b}(g_1,g_2)}\cdot\\ &\quad\cdot(z_2-z_3)^{\hat{b}(g_2,g_3)}(z_2-z_3+z_0)^{\hat{b}(g_1,g_3)})\\ &= \Omega(g_1+g_2,g_3)\iota_{32}^{-1}(\langle v', Y(v_3,z_3)Y(Y(v_1,z_0)v_2,z_2)\mathbf{1}\rangle\cdot\\ &\quad\cdot z_0^{\hat{b}(g_1,g_2)}(e^{-i\pi}(z_3-z_2))^{\hat{b}(g_2,g_3)}(e^{-i\pi}(z_3-z_2)+z_0)^{\hat{b}(g_1,g_3)})\\ &= \Omega(g_1+g_2,g_3)\iota_{32}^{-1}(\langle v', Y(v_3,z_3)Y(v_1,z_0+z_2)Y(v_2,z_2)\mathbf{1}\rangle\cdot\\ &\quad\cdot z_0^{\hat{b}(g_1,g_2)}(e^{-i\pi}(z_3-z_2))^{\hat{b}(g_2,g_3)}(e^{-i\pi}(z_3-z_0-z_2))^{\hat{b}(g_1,g_3)})\\ &= \Omega(g_1+g_2,g_3)\Omega(g_2,g_3)^{-1}B(g_1,g_3,g_2)^{-1}\cdot\\ (12.140)\qquad &\quad\cdot\iota_{32}^{-1}(\iota_{312}f(z_1,z_2,z_3)|_{z_1=z_0+z_2})\\ &= F(g_1,g_2,g_3)^{-1}\iota_{32}^{-1}(\iota_{312}f(z_1,z_2,z_3)|_{z_1=z_0+z_2})\\ &= F(g_1,g_2,g_3)^{-1}\iota_{32}^{-1}(g(z_1,z_2,z_3)\iota_{12}((z_1-z_2)^{-p})\cdot\\ &\quad\cdot\iota_{31}((z_1-z_3)^{-q})\iota_{32}((z_2-z_3)^{-r})|_{z_1=z_0+z_2})\\ &= F(g_1,g_2,g_3)^{-1}\iota_{32}^{-1}(g(z_0+z_2,z_2,z_3)z_0^{-p}\cdot\\ &\quad\cdot\iota_{320}((z_0+z_2-z_3)^{-q})\iota_{32}((z_2-z_3)^{-r}))\\ &= F(g_1,g_2,g_3)^{-1}g(z_0+z_2,z_2,z_3)z_0^{-p}\cdot\\ &\quad\cdot\iota_{40}((z_0+z_4)^{-q})z_4^{-r}|_{z_4=z_2-z_3}\end{aligned}$$

(the identities (12.102) and (12.111) are used here), so that

$$F(g_1,g_2,g_3)\langle v', Y(Y(v_1,z_0)v_2,z_2)Y(v_3,z_3)\mathbf{1}\rangle\cdot$$

$$\cdot z_0^{\hat{b}(g_1,g_2)}(z_2 - z_3)^{\hat{b}(g_2,g_3)}(z_2 - z_3 + z_0)^{\hat{b}(g_1,g_3)}$$
$$= \iota_{230}(g(z_0 + z_2, z_2, z_3)z_0^{-p}(z_0 + z_2 - z_3)^{-q}(z_2 - z_3)^{-r})$$
$$= \iota_{230}f(z_0 + z_2, z_2, z_3). \tag{12.141}$$

Exactly as in (7.54)-(7.56), we finally obtain

$$\langle v', Y(v_1, z_0 + z_2)Y(v_2, z_2)Y(v_3, z_3)\mathbf{1}\rangle \cdot$$
$$\cdot(z_0 + z_2 - z_3)^{\hat{b}(g_1,g_3)}(z_2 - z_3)^{\hat{b}(g_2,g_3)}z_0^{\hat{b}(g_1,g_2)}$$
$$= (\iota_{123}f(z_1, z_2, z_3))|_{z_1=z_0+z_2} \tag{12.142}$$
$$= \iota_{023}f(z_0 + z_2, z_2, z_3),$$

$$F(g_1, g_2, g_3)\langle v', Y(Y(v_1, z_0)v_2, z_2)v_3\rangle(z_2 + z_0)^{\hat{b}(g_1,g_3)}\cdot$$
$$\cdot z_2^{\hat{b}(g_2,g_3)}z_0^{\hat{b}(g_1,g_2)} = \iota_{20}f(z_0 + z_2, z_2, 0), \tag{12.143}$$

$$\langle v', Y(v_1, z_0 + z_2)Y(v_2, z_2)v_3\rangle(z_0 + z_2)^{\hat{b}(g_1,g_3)}\cdot$$
$$\cdot z_2^{\hat{b}(g_2,g_3)}z_0^{\hat{b}(g_1,g_2)} = \iota_{02}f(z_0 + z_2, z_2, 0), \tag{12.144}$$

as desired. □

Remark 12.36: The results of Chapter 8 on monodromy and one-dimensional braid group representations carry over with only the following modifications (using the setting of Theorem 12.33): First, the expression (8.11), with $(i_1 \cdots i_n)$ taken to be the transposition of i and $i + 1$, is multiplied by the extra factor $e^{-i\pi\hat{b}(g_i,g_{i+1})}B(g_i, g_{i+1}, g_{i+2} + \cdots + g_{n+1})$ occurring in (12.136). Next, the factor $(e^{2\pi i/2T})^{(g_i,g_{i+1})\epsilon T} = e^{i\pi(g_i,g_{i+1})\epsilon}$ in the right-hand side of (8.13), with $\epsilon = -$, is replaced by $B(g_i, g_{i+1}, g_{i+2} + \cdots + g_{n+1})^{-1}$. Note that in Chapter 8, in formula (8.18), we made an arbitrary choice of orientation in our realization of the braid group; the factor $e^{i\pi t}$ in (8.18) could alternatively be replaced by $e^{-i\pi t}$, which would entail the change from $\epsilon = +$ to $\epsilon = -$ starting at (8.21). However, we have made an analogous arbitrary choice in the present chapter, a choice reflected for instance in our use of the expression $\frac{z_2 - z_1}{e^{i\pi}z_0}$ rather than $\frac{z_2 - z_1}{e^{-i\pi}z_0}$ in the second term of the left-hand side of the Jacobi identity. It turns out that in the context of the present chapter, the opposite choice of realization of the braid group in (8.18), and correspondingly, the opposite choice of ϵ (namely, $\epsilon = -$) becomes appropriate, as we have just seen. Taking $\epsilon = -$ changes the sign in the exponents in the right-hand sides of (8.21) and (8.22). We now see that in the present setting, the exponential factor in the right-hand sides of (8.21) and (8.22) is replaced by $B(g_j, g_{j+1}, g_{j+2} + \cdots + g_n)$. When each g_i equals g, as in (8.24), this quantity becomes $\Omega(g, g) = q(g)$.

We have:

Theorem 12.37 *Under the hypothesis (9.39), the assertions of Proposition 8.3 and of Remark 8.4 remain valid, with the braid group B_n acting according to the quadratic form $q(\cdot)$:*

$$b_k \mapsto q(g) \quad \text{for} \quad k = 1, ..., n-1, \tag{12.145}$$

and with the Hecke algebra parameter q (see Remark 8.4) taken to be $q(g)$. □

Remark 12.38: We continue our discussion of vertex superalgebras from Remark 9.21. Since the notion of abelian intertwining algebra generalizes that of generalized vertex algebra defined in Chapter 9 (recall Remark 12.28), vertex superalgebras are in fact special cases of abelian intertwining algebras. The structures discussed at the beginning of this chapter provide an important family of examples of vertex superalgebras – those based canonically on integral lattices (for instance, weight lattices of type B_l): Let L be an integral lattice (i.e., $\langle L, L\rangle \subset \mathbf{Z}$). Then L is the union of an even sublattice L_0, consisting exactly of all the elements α such that $\langle \alpha, \alpha\rangle \in 2\mathbf{Z}$, and (at most) one coset L_1, consisting of all the elements α such that $\langle \alpha, \alpha\rangle \in 2\mathbf{Z}+1$. The formula

$$c(\alpha, \beta) = (-1)^{\langle \alpha,\beta\rangle + \langle \alpha,\alpha\rangle\langle \beta,\beta\rangle}$$

for $\alpha, \beta \in L$ defines an alternating bilinear map from $L \times L$ to $\langle \pm 1\rangle$, and the Jacobi identity (12.19) reduces to the super-Jacobi identity (6.66), where we provide V_L with the obvious $\mathbf{Z}/2\mathbf{Z}$-grading determined by L_0 and L_1. (The modification (12.3) amounts precisely to a sign change of the operators parametrized by the vectors in the "odd" subspace of V_L and mapping this subspace to the "even" subspace.) We thus obtain a vertex superalgebra from Theorem 12.24, where we take L, L_0 and c as above, with L not an even lattice (that is, $L \neq L_0$), $T = 2$ and the function f of (12.90) to be identically 1. Then we also have $G = \mathbf{Z}/2\mathbf{Z}$; the function F, which is the function h of (12.20), is identically 1; the function Ω (recall (12.100)) is given by $\Omega(g_1, g_2) = (-1)^{g_1 g_2}$ for g_1, $g_2 \in \mathbf{Z}/2\mathbf{Z}$; the quadratic form q (recall (12.103)) is given by $q(g) = (-1)^g$ for $g \in \mathbf{Z}/2\mathbf{Z}$; the form $\hat{b}$ (recall Remark 12.28) is given by multiplication in $\mathbf{Z}/2\mathbf{Z}$; and we obtain a vertex superalgebra $(V, Y, \mathbf{1}, \omega, \mathbf{Z}/2\mathbf{Z})$.

Just as in Chapter 10, it is easy to form tensor products of abelian intertwining algebras: Let

$$V_i = (V_i, Y_i, \mathbf{1}_i, \omega_i, T_i, G_i, F_i, \Omega_i), \qquad i = 1, ..., n \tag{12.146}$$

be a finite family of abelian intertwining algebras. Set

$$T = \text{l.c.m.}(T_1, ..., T_n), \tag{12.147}$$

$$G = G_1 \times \cdots \times G_n, \tag{12.148}$$

$$F = F_1 \times \cdots \times F_n \tag{12.149}$$

$$\Omega = \Omega_1 \times \cdots \times \Omega_n \tag{12.150}$$

(in the obvious senses). Then F: $G \times G \times G \to \mathbf{C}^{\times}$ and Ω: $G \times G \to \mathbf{C}^{\times}$ satisfy the appropriate conditions (including (12.118)). We have the vector space

$$V = V_1 \otimes \cdots \otimes V_n \tag{12.151}$$

equipped with the two compatible gradations $V = \coprod_{n \in \frac{1}{T}\mathbf{Z}} V_n$ and $V = \coprod_{g \in G} V^g$ and with the operators

$$Y(v_1 \otimes \cdots \otimes v_n, z) = Y_1(v_1, z) \otimes \cdots \otimes Y_n(v_n, z) \tag{12.152}$$

and the elements $\mathbf{1}$ and ω, as in (10.4)-(10.8), so that V is a tensor product of Virasoro algebra modules. Using the duality result Theorem 12.32 for proving the Jacobi identity, we see easily that the analogue of Proposition 10.1 holds:

Proposition 12.39 *The tensor product $V = (V, Y, \mathbf{1}, \omega, T, G, F, \Omega)$ is an abelian intertwining algebra with rank equal to the sum of the ranks of the V_i.* □

Chapter 13

Affine Lie algebras and vertex operator algebras

In this chapter we shall discuss standard (or integrable highest weight) representations of affine Kac-Moody algebras from the point of view of vertex operator algebra theory. This will also provide the setting for the next chapter. We restrict our attention to a simple Lie algebra $\mathcal{G}$ of type A, D or E. Let $\hat{\mathcal{G}}$ be the corresponding affine Lie algebra and let l be a positive integer. We show that a certain distinguished one of the level l standard $\hat{\mathcal{G}}$-modules $L(l,0)$ has the structure of a vertex operator algebra and that every level l standard $\hat{\mathcal{G}}$-module, at least every such module which can be obtained from the tensor product of l basic modules (i.e., level 1 standard modules for affine Lie algebras of type $\hat{A}$, $\hat{D}$ or $\hat{E}$), is an irreducible module for this vertex operator algebra. Conversely, every irreducible module for the vertex operator algebra $L(l,0)$ is also a standard $\hat{\mathcal{G}}$-module of level l. The Virasoro algebra for the vertex operator algebra $L(l,0)$ comes from a canonical vertex operator associated with the (suitably normalized) Casimir operator for $\mathcal{G}$. Using an abelian intertwining algebra (see Chapter 12) associated with the direct sum of l copies of the weight lattice, we also construct intertwining operators for irreducible $L(l,0)$-modules. There is some overlap between the results in this chapter and those in [FZ], which uses a different approach, involving Verma modules. It is also possible to study the higher level standard modules for affine algebras by means of the "full subalgebra" approach instead of the tensor product approach, as in [LP2], for example, and it would be interesting to extend this approach to vertex operator algebras. (Given a $\mathbf{Z}$-graded Lie algebra and a positive integer N, the *full subalgebra of depth* N is defined to be the graded subalgebra which is the direct sum of the homogeneous subspaces whose degrees are divisible by N.) For background material for this chapter, see especially [FLM3], [K] and [L1]. The material in this chapter provides an algebraic foundation for the Wess-Zumino-Novikov-Witten model in conformal field theory (cf. [GW], [KZ], [W1], [W2]). Our treatment should be compared with the (mathematical) treatment of related matters in [TK]. Here we are making the vertex operator algebra and module structure explicit.

After introducing the underlying finite-dimensional simple Lie algebra $\mathcal{G}$ of type A, D or E and its affinization, we recall the untwisted vertex operator construction of the level 1 standard $\hat{\mathcal{G}}$-modules ([FK], [Se]; see also [Ha], [BHN]). We then consider the (usually reducible) level l $\hat{\mathcal{G}}$-module constructed from the diagonal action of the affine Lie algebra on an untwisted space V_L formed from the direct sum L of l copies of the weight lattice of $\mathcal{G}$, and we thus focus our attention on the level l standard $\hat{\mathcal{G}}$-modules which can be embedded as irreducible components of tensor products of l basic modules. The standard module $L(l,0)$ is a distinguished one of these modules. Using the Casimir element of $\mathcal{G}$, we construct a canonical quadratic element of V_L whose corresponding vertex operator turns out to generate the "Segal-Sugawara" realization of the Virasoro algebra (see [Su], [BH], [Se]). This is accomplished with the help of a simple but useful general principle (Proposition 13.4) which, in the context of an arbitrary vertex operator algebra and module, indicates (by means of the Jacobi identity) that the vertex operator associated with a symmetrized product of a pair of weight-one elements can be expressed as a normal ordered product of two vertex operators, each generated by one of the weight-one vectors. The normal-ordered quadratic nature of the Virasoro algebra operators is a natural and immediate consequence. For the derivation of the resulting bracket action of the Virasoro algebra on the affine Lie algebra, we have made use of a simplification, supplied by Haisheng Li, of our earlier argument. The vertex operators associated with the affine Lie algebra are primary fields of weight 1 with respect to the Virasoro algebra. The relationship between this Virasoro algebra and the one associated naturally with V_L (see Proposition 13.8) is closely related to a special case of the "coset construction" of [GKO]. We show that the two Virasoro algebras agree when $l = 1$ (Proposition 13.10), giving an equality between a canonical expression built from an orthonormal basis of $\mathcal{G}$ and a different canonical expression built from an orthonormal basis of its Cartan subalgebra. All these results are summarized in a theorem describing the structure of $L(l,0)$ as a simple (i.e., irreducible as a module for itself) vertex operator algebra generated by its subspace of weight 1, which is a canonical copy of $\mathcal{G}$.

Next we turn to the decomposition of V_L as a direct sum of irreducible $\hat{\mathcal{G}}$-modules isomorphic to standard modules of level l, which we denote $L(l,\lambda)$, the elements λ being the highest weights of certain $\mathcal{G}$-modules (generalizing the case $\lambda = 0$ that we have been considering). We show that after we perform a suitable grading-shift, each of these standard modules gains the structure of an irreducible module for the vertex operator algebra $L(l,0)$. Then we prove conversely that every irreducible module for the vertex operator algebra $L(l,0)$ carries the structure of a standard $\hat{\mathcal{G}}$-module of level l. This is accomplished with the help of another simple but useful general principle about vertex operator algebras and modules (Proposition 13.16), which, under the assumption that the component operators of a vertex operator $Y(v,z)$ commute with one another, express each power of this operator $Y(v,z)$ as the vertex operator of a certain vector. Then we construct canonical intertwining operators among the irreducible $L(l,0)$-modules by suitably modifying the underlying vertex operator $Y(v,z)$ for the ambient space V_L in two stages: First we modify the operators so as to make V_L an abelian intertwining algebra as in Theorem 12.24, and then we further modify the operators in a way which takes into account the grading-shift mentioned above for the modules $L(l,\lambda)$. In this way we construct our family of intertwining operators among the modules $L(l,\lambda)$.

Let $\mathcal{G}$ be a finite-dimensional simple Lie algebra of type A, D or E with nonsingular invariant symmetric bilinear form $\langle\cdot,\cdot\rangle$, let $\mathcal{H}$ be a Cartan subalgebra, let Δ be the root system of $\mathcal{G}$, viewed as a subset of $\mathcal{H}$, and assume that the form is normalized so that $\langle\alpha,\alpha\rangle = 2$ for $\alpha \in \Delta$. Denote by Q and P the root lattice and weight lattice, respectively, understood as embedded in $\mathcal{H}$, so that Q is an even sublattice of P. Recalling the setting of Chapters 2 and 3, take $L = P$ and a central extension $\hat{P}$ of P by $\langle\kappa_q\rangle$, where q is even, with commutator map c_0, and with $c(\alpha,\beta) = \omega_q^{c_0(\alpha,\beta)}$ for $\alpha,\beta \in P$. We assume that $c(\alpha,\beta) = (-1)^{\langle\alpha,\beta\rangle}$ if $\alpha,\beta \in Q$.

We choose a section

$$
\begin{aligned}
e : P &\to \hat{P} \\
\alpha &\mapsto e_\alpha,
\end{aligned}
\tag{13.1}
$$

and we denote by ϵ_0: $P \times P \to \mathbf{Z}/q\mathbf{Z}$ the corresponding 2-cocycle.

Then there exist $x_\alpha \in \mathcal{G}$ $(\alpha \in \Delta)$ such that

$$\mathcal{G} = \mathcal{H} \oplus \sum_{\alpha \in \Delta} \mathbf{C} x_\alpha, \tag{13.2}$$

with

$$[\mathcal{H}, \mathcal{H}] = 0 \tag{13.3}$$

$$[h, x_\alpha] = \langle h, \alpha \rangle x_\alpha \tag{13.4}$$

$$[x_\alpha, x_\beta] = \begin{cases} \epsilon(\alpha, -\alpha)\alpha & \text{if } \alpha + \beta = 0 \\ \epsilon(\alpha, \beta) x_{\alpha+\beta} & \text{if } \alpha + \beta \in \Delta \\ 0 & \text{if } \alpha + \beta \notin \Delta \cup \{0\} \end{cases} \tag{13.5}$$

for $h \in \mathcal{H}$ and $\alpha, \beta \in \Delta$, where we set

$$\epsilon(\alpha, \beta) = \omega_q^{\epsilon_0(\alpha, \beta)} \quad \text{for} \quad \alpha, \beta \in P. \tag{13.6}$$

The bilinear form $\langle \cdot, \cdot \rangle$ on $\mathcal{G}$ satisfies the conditions

$$\langle \mathcal{H}, x_\alpha \rangle = 0 \quad \text{for} \quad \alpha \in \Delta \tag{13.7}$$

$$\langle x_\alpha, x_\beta \rangle = \begin{cases} \epsilon(\alpha, -\alpha) & \text{if } \ \alpha + \beta = 0 \\ 0 & \text{if } \ \alpha + \beta \neq 0 \end{cases} \tag{13.8}$$

for $\alpha, \beta \in \Delta$.

The affine Lie algebra $\hat{\mathcal{G}}$ is the infinite-dimensional Lie algebra

$$\hat{\mathcal{G}} = \mathcal{G} \otimes \mathbf{C}[t, t^{-1}] \oplus \mathbf{C}c \tag{13.9}$$

with the bracket relations

$$[x \otimes t^m, y \otimes t^n] = [x, y] \otimes t^{m+n} + \langle x, y \rangle m \delta_{m+n,0} c, \tag{13.10}$$

where $x, y \in \mathcal{G}$, $m, n \in \mathbf{Z}$ and c is a central element. For $x \in \mathcal{G}$, set

$$x(z) = \sum_{n \in \mathbf{Z}} (x \otimes t^n) z^{-n-1}. \tag{13.11}$$

From now on we fix a positive integer l. Set

$$L = P_1 \oplus \cdots \oplus P_l \tag{13.12}$$

where each P_i is a copy of P. We write $\alpha_i \in P_i$ for the element corresponding to $\alpha \in P$, and we use analogous notation for subsets of P (for example, Q). We extend the form $\langle\cdot,\cdot\rangle$ to L so that P_i and P_j are orthogonal if $i \neq j$. Set

$$L_0 = Q_1 \oplus \cdots \oplus Q_l \subset L. \tag{13.13}$$

Then L_0 is the dual lattice of L in

$$\mathbf{h} = \mathbf{C} \otimes_{\mathbf{Z}} L. \tag{13.14}$$

Let $\hat{L}$ be the central extension of L by $\langle\kappa_q\rangle$ with commutator map which is the sum, in the obvious sense, of the commutator maps on the P_i. Note that the sum of the 2-cocycles on the P_i is a 2-cocycle on L associated with $\hat{L}$. We again denote these by $c_0(\cdot,\cdot)$ and $\epsilon_0(\cdot,\cdot)$, respectively, without confusion. Then ϵ_0 is associated with a section $e : L \to \hat{L}$, compatible with (13.1). We have $c(\alpha,\beta) = (-1)^{\langle\alpha,\beta\rangle}$ for $\alpha,\beta \in L_0$ (where $c(\cdot,\cdot)$ is defined as usual), and by the results in Chapter 5 or 6 or Remark 9.10, V_{L_0} is a vertex operator algebra (since L is positive definite) and V_L is a V_{L_0}-module.

We introduce a linear injection i from $\mathcal{G}$ to V_{L_0} as follows:

$$i(h) = h_1(-1) + \cdots + h_l(-1) \in V_{L_0} \quad \text{for} \quad h \in \mathcal{H} \tag{13.15}$$

$$i(x_\alpha) = \iota(e_{\alpha_1}) + \cdots + \iota(e_{\alpha_l}) \in V_{L_0} \quad \text{for} \quad \alpha \in \Delta \tag{13.16}$$

(recall the notations $h(-1)$ and $\iota(a)$ from Chapter 3). Recall that the *level* of a $\hat{\mathcal{G}}$-module W is $k \in \mathbf{C}$ if c acts on W as multiplication by k.

Proposition 13.1 *The linear map* $\pi : \hat{\mathcal{G}} \to \operatorname{End} V_L$ *given by*

$$\begin{aligned} \pi(c) &= l \\ \pi(x(z)) &= Y(i(x), z) \quad \text{for} \quad x \in \mathcal{G} \end{aligned} \tag{13.17}$$

defines a $\hat{\mathcal{G}}$*-module structure on* V_L*, of level* l*. Moreover,* V_L *is completely reducible under* $\hat{\mathcal{G}}$ *and every irreducible component is a level* l *standard* $\hat{\mathcal{G}}$*-module.* □

This follows from two remarks:

Remark 13.2: For $l = 1$, see Section 7.2 of [FLM3]; in this case, V_L is a direct sum of one copy of each of the basic $\hat{\mathcal{G}}$-modules (the standard level 1 $\hat{\mathcal{G}}$-modules).

Remark 13.3: For general l, as $\hat{\mathcal{G}}$-module, V_L is isomorphic to the tensor product of l copies of V_P.

For $x \in \mathcal{G}$ and $n \in \mathbf{Z}$, we write $x(n)$ for the operator $\pi(x \otimes t^n)$ on V_L. Then the linear map i can be written as $i(x) = x(-1) \cdot \iota(1)$, by the creation property (3.45), and we view $\mathcal{G}$ as embedded in V_{L_0} accordingly, as a subspace of elements of weight 1. By comparing the expression (13.11) with the generic expansion $Y(x, z) = \sum_{n\in\mathbf{Z}} x_n z^{-n-1}$, we see that

$$x_n = x(n). \tag{13.18}$$

These operators satisfy the bracket relations corresponding to (13.10), with $c \mapsto l$, on V_L.

Set

$$L(l, 0) = U(\hat{\mathcal{G}}) \cdot \iota(1) \, (\supset \mathcal{G}). \tag{13.19}$$

Then $L(l, 0) \subset V_{L_0}$ is a standard (irreducible) $\hat{\mathcal{G}}$-module. Moreover, in view of [FHL], (2.4.6), $L(l, 0)$ is exactly the "vertex operator subalgebra without Virasoro algebra" (in the obvious sense) of V_{L_0} generated by $i(\mathcal{G})$. However, as will become clear later, $\omega \notin L(l, 0)$ if $l > 1$ (recall (5.65)). Our next goal is to find another Virasoro element $\omega_{\mathcal{G}_l}$ of V_{L_0} which lies in $L(l, 0)$ such that $L(l, 0)$ will become a vertex operator algebra. We shall take $\omega_{\mathcal{G}_l}$ to be a suitably normalized canonical "quadratic" $\mathcal{G}$-invariant element.

Let $d = \dim \mathcal{G}$ and let $\{u_1, ..., u_d\}$ be an orthonormal basis of $\mathcal{G}$ with respect to the form $\langle \cdot, \cdot \rangle$ (see (13.7)-(13.8)) and let h be the Coxeter number of $\mathcal{G}$. Then the Casimir element of $U(\mathcal{G})$ (which acts as the identity operator on $\mathcal{G}$ under the adjoint representation) is given by

$$\Omega = \frac{1}{2h} \sum_{i=1}^{d} u_i^2, \tag{13.20}$$

or equivalently, the Killing form of $\mathcal{G}$ is $2h$ times the form $\langle \cdot, \cdot \rangle$, since $\sum_{\alpha\in\Delta} \langle \alpha, \beta \rangle^2 = 2h\langle \beta, \beta \rangle$ for $\beta \in \mathcal{H}$ (cf. [Bou]). Set

$$\omega_{\mathcal{G}_l} = \frac{h}{l+h} \Omega = \frac{1}{2(l+h)} \sum_{i=1}^{d} u_i(-1)^2 \iota(1) \in L(l, 0), \tag{13.21}$$

where we identify Ω with an element of $L(l, 0)$ as indicated. (Note that formally, $\omega_{\mathcal{G}_l} = \Omega$ for $l = 0$, which is a degenerate situation.) We shall show below that $Y(\omega_{\mathcal{G}_l}, z)$ gives the right Virasoro algebra, in the

Segal-Sugawara form (cf. [Se], [Su]). The reader will see that this form of the operators (and the properties of these operators) are canonically determined by the Jacobi identity from the Casimir element of $\mathcal{G}$ in a direct conceptual way.

For this, we briefly digress to explain in general how certain "quadratic" elements in a vertex operator algebra naturally give rise to normal ordered products. Let V be any vertex operator algebra and let W be any V-module, with the corresponding vertex operators denoted Y_V and Y_W, respectively. Let $u, v \in V$. Then Res_{z_1} of the Jacobi identity acting on W gives:

$$
\begin{aligned}
Y_W(Y_V(u,z_0)v,z_2) &= \mathrm{Res}_{z_1} z_0^{-1}\delta\left(\frac{z_1-z_2}{z_0}\right) Y_W(u,z_1)Y_W(v,z_2) \\
&\quad -\mathrm{Res}_{z_1} z_0^{-1}\delta\left(\frac{z_2-z_1}{-z_0}\right) Y_W(v,z_2)Y_W(u,z_1).
\end{aligned}
\tag{13.22}
$$

(Sometimes we "simplify" the first term on the right to $Y_W(u, z_0 + z_2)Y_W(v, z_2)$, as in (7.9), but now we want both terms on the right to look similar.) If we further take Res_{z_0}, we get

$$
Y_W(u_0 v, z_2) = [u_0, Y_W(v, z_2)].
\tag{13.23}
$$

But instead we take $\mathrm{Res}_{z_0} z_0^{-1}\cdot$ (i.e., we take the coefficient of z_0^0) to get

$$
\begin{aligned}
Y_W(u_{-1}v,z_2) &= \mathrm{Res}_{z_1}(z_1-z_2)^{-1}Y_W(u,z_1)Y_W(v,z_2) \\
&\quad + \mathrm{Res}_{z_1}(z_2-z_1)^{-1}Y_W(v,z_2)Y_W(u,z_1) \\
&= \sum_{m<0} u_m z_2^{-m-1} Y_W(v,z_2) + Y_W(v,z_2)\sum_{m\geq 0} u_m z_2^{-m-1}.
\end{aligned}
\tag{13.24}
$$

This suggests defining a "normal ordering" operation by: For $u, v \in V$, $m, n \in \mathbf{Z}$,

$$
{}^{\times}_{\times} u_m v_n {}^{\times}_{\times} = \begin{cases} u_m v_n & \text{if } m < 0 \\ v_n u_m & \text{if } m \geq 0. \end{cases}
\tag{13.25}
$$

Then (since $v = v_{-1}\mathbf{1}$)

$$
Y_W(u_{-1}v_{-1}\mathbf{1}, z_2) = Y_W(u_{-1}v, z_2) = {}^{\times}_{\times} Y_W(u,z_2)Y_W(v,z_2) {}^{\times}_{\times}.
\tag{13.26}
$$

However, this normal ordering is not in general a commutative operation, since if m and n are both negative or both nonnegative, ${}^{\times}_{\times} u_m v_n {}^{\times}_{\times}$ and ${}^{\times}_{\times} v_n u_m {}^{\times}_{\times}$ differ by $\pm[u_m, v_n]$. (But note that (13.26) applies to arbitrary u and v, not necessarily of weight 1.)

Formula (13.26) may be applied iteratively; see for instance Proposition 13.16 below.

Assume now that

$$\text{wt}\, u = \text{wt}\, v = 1, \tag{13.27}$$

so that

$$\text{wt}\, u_n = \text{wt}\, v_n = -n.$$

Also consider the normal ordering operation given by:

$$:u_m v_n: = \begin{cases} u_m v_n & \text{if} \quad m < n \\ \frac{1}{2}(u_m v_n + v_n u_m) & \text{if} \quad m = n \\ v_n u_m & \text{if} \quad m > n \end{cases} \tag{13.28}$$

This is a commutative operation. (Compare this with the normal ordering operation $\overset{\circ}{\circ}\cdot\overset{\circ}{\circ}$ of Chapter 3.)

Assume in addition that for all u and v of weight 1,

$$u_1 v \in \mathbf{C}\mathbf{1}, \quad u_n v = 0 \quad \text{if} \quad n > 1 \tag{13.29}$$

(the usual affine Lie algebra situation, as above). Then by the Jacobi identity, we have the affine Lie algebra relations

$$[u_m, v_n] = (u_0 v)_{m+n} + (u_1 v) m \delta_{m+n,0}, \tag{13.30}$$

where $u_1 v$ is identified with a scalar operator on W. (Note that the comparison of these relations with (13.10) allows us to identify $[x, y]$ with $x_0 y$ and $l\langle x, y\rangle$ with $x_1 y$ for $x, y \in \mathcal{G}$, or rather, $x, y \in i(\mathcal{G}) \subset V_{L_0}$, in our special situation.)

Proposition 13.4 *Let* u, v *be as indicated. Then*

$$\begin{aligned} Y_W(u_{-1}v_{-1}\mathbf{1} + v_{-1}u_{-1}\mathbf{1}, z) &= Y_W(u_{-1}v + v_{-1}u, z) \\ &= {}^{\times}_{\times} Y_W(u,z) Y_W(v,z) {}^{\times}_{\times} + {}^{\times}_{\times} Y_W(v,z) Y_W(u,z) {}^{\times}_{\times} \\ &= 2{:}Y_W(u,z) Y_W(v,z){:}. \end{aligned} \tag{13.31}$$

In particular,

$$Y_W((u_{-1})^2 \mathbf{1}, z) = {}^{\times}_{\times} Y_W(u,z)^2 {}^{\times}_{\times} = {:}Y_W(u,z)^2{:}. \tag{13.32}$$

Proof Since (13.31) follows from (13.32) by polarization we prove the latter. We know that $u_0u = 0$, from (13.30) (for instance, $m = 0$, $n = -1$ gives $(u_0u)_{-1} = 0$; now apply to $\mathbf{1}$), so that u_m and u_n commute if $m \neq -n$. Thus

$$\text{(13.33)} \qquad {}^{\times}_{\times} u_m u_n {}^{\times}_{\times} = {:}u_m u_n{:} \quad \text{if} \quad m \neq -n$$

and the same relation clearly holds if $m = -n$. The proof is complete by (13.26). □

Now we return to the space $L(l, 0)$. From (13.18), (13.21) and (13.32), applied to V_{L_0}-module V_L, we find, setting

$$\text{(13.34)} \qquad Y(\omega_{\mathcal{G}_l}, z) = L_{\mathcal{G}_l}(z) = \sum_{k \in \mathbf{Z}} L_{\mathcal{G}_l}(k) z^{-k-2},$$

that

$$\text{(13.35)} \qquad L_{\mathcal{G}_l}(k) = \frac{1}{2(l+h)} \sum_{i=1}^{d} \sum_{m \in \mathbf{Z}} {:}u_i(m) u_i(k-m){:}$$

as operators on V_L. In fact, the argument has shown that if W is any module (in the obvious sense) for the "vertex operator algebra without Virasoro algebra" $L(l, 0)$, then $Y(\omega_{\mathcal{G}_l}, z)$ acts on W by means of the operators (13.35). These are the so-called Segal-Sugawara operators (cf. [Se], [Su]), and we have shown that they are the components of a vertex operator. It is easy to derive their properties using the Jacobi identity:

Proposition 13.5 *For $x \in \mathcal{G}$ and $m, n \in \mathbf{Z}$,*

$$\text{(13.36)} \qquad [L_{\mathcal{G}_l}(m), x(n)] = -n x(m+n)$$

on V_L.

Proof Clearly, $x(m)\omega_{\mathcal{G}_l} = 0$ if $m > 2$ because $\operatorname{wt} x(m)\omega_{\mathcal{G}_l} < 0$. Now we compute $x(2)\omega_{\mathcal{G}_l}$, using the affine Lie algebra relations (13.10), with c acting as l:

$$\begin{aligned} x(2)\omega_{\mathcal{G}_l} &= \frac{1}{2(l+h)} \sum_{i=1}^{d} x(2) u_i(-1)^2 \iota(1) \\ &= \frac{1}{2(l+h)} \sum_{i=1}^{d} ([x, u_i](1) u_i(-1)\iota(1) + u_i(-1) x(2) u_i(-1)\iota(1)) \end{aligned}$$

$$\begin{aligned} &= \frac{1}{2(l+h)} \sum_{i=1}^{d} l\langle [x, u_i], u_i\rangle \iota(1) \\ &= \frac{1}{2(l+h)} \sum_{i=1}^{d} l\langle x, [u_i, u_i]\rangle \iota(1) \\ &= 0, \end{aligned}$$

where we use the fact that $x(m)\iota(1) = 0$ for $x \in \mathcal{G}$ and $m \geq 0$. Similarly,

$$\begin{aligned} x(1)\omega_{\mathcal{G}_l} &= \frac{1}{2(l+h)} \sum_{i=1}^{d} ([x, u_i](0)u_i(-1)\iota(1) + 2l\langle x, u_i\rangle u_i(-1)\iota(1)) \\ &= \frac{1}{2(l+h)} \left(\sum_{i=1}^{d} [[x, u_i], u_i](-1)\iota(1) + 2lx(-1)\iota(1) \right) \\ &= x(-1)\iota(1), \end{aligned}$$

since Ω acts as the identity operator on $\mathcal{G}$ (recall (13.20)), and

$$x(0)\omega_{\mathcal{G}_l} = 0,$$

by applying the $\mathcal{G}$-module map

$$\begin{aligned} \mathcal{G} \otimes \mathcal{G} &\to V_L \\ u \otimes v &\mapsto u(-1)v(-1)\iota(1) \end{aligned}$$

to the relation $x \cdot \sum u_i \otimes u_i = 0$. That is,

(13.37) $\quad x(z)\omega_{\mathcal{G}_l} = x(-1)\iota(1)z^{-2} + \text{a nonsingular series in } z,$

where $x(z)$ is the vertex operator

$$x(z) = Y(x(-1)\iota(1), z).$$

Thus the Jacobi identity gives

$$[x(z_1), L_{\mathcal{G}_l}(z_2)] = -z_2^{-1}x(z_2)\frac{\partial}{\partial z_1}\delta(z_1/z_2) \tag{13.38}$$

or equivalently,

$$[x(m), L_{\mathcal{G}_l}(n)] = mx(m+n),$$

as desired. □

Remark 13.6: Formula (13.36) amounts to the condition that $x(z)$ is a primary field of weight 1 with respect to the Virasoro algebra generated by the $L_{\mathcal{G}_l}(m)$ – once we have finished proving the Virasoro algebra properties, of course (cf. Proposition 5.9). A similar comment applies to (13.39) below.

Recalling the original Virasoro operators $L(m)$, $m \in \mathbf{Z}$, for V_L and the relations

$$[L(m), x(n)] = -nx(m+n) \tag{13.39}$$

for $x \in \mathcal{G}$ and $m, n \in \mathbf{Z}$ (from the primary field relations (5.81), for instance), we find that

$$[Y(\omega-\omega_{\mathcal{G}_l}, z_1), x(z_2)] = [Y(\omega, z_1) - Y(\omega_{\mathcal{G}_l}, z_1), x(z_2)] = 0 \text{ on } V_L. \tag{13.40}$$

Remark 13.7: This is a special case of a result in the "coset construction," in the sense of [GKO], in this case, for the pair ($\mathcal{G} \times \cdots \times \mathcal{G}, \mathcal{G}$ (embedded diagonally)).

Proposition 13.8 *For $m \geq -1$,*

$$L(m) = L_{\mathcal{G}_l}(m) \tag{13.41}$$

on $L(l,0)$. In particular, $L_{\mathcal{G}_l}(0)$ defines the given grading of $L(l,0)$, and for $v \in L(l,0)$,

$$Y(L_{\mathcal{G}_l}(-1)v, z) = \frac{d}{dz} Y(v, z) \tag{13.42}$$

on V_L.

Proof The second assertion is a direct consequence of the first together with the $L(0)$-grading property and the $L(-1)$-property (5.71) for V_L. By (13.40), it is enough to show that

$$L(m)\iota(1) = L_{\mathcal{G}_l}(m)\iota(1) = 0 \tag{13.43}$$

for $m \geq -1$, which is obvious for $m > 0$ and for $L(0)$ and $L(-1)$. The rest follows from (13.35) and the fact that $x(0)\iota(1) = 0$ for $x \in \mathcal{G}$. □

Proposition 13.9 *On V_L,*

$$\begin{aligned}[L_{\mathcal{G}_l}(z_1), L_{\mathcal{G}_l}(z_2)] &= z_2^{-1}\left(\frac{d}{dz_2} L_{\mathcal{G}_l}(z_2)\right)\delta(z_1/z_2) \\ &-2z_2^{-1} L_{\mathcal{G}_l}(z_2)\frac{\partial}{\partial z_1}\delta(z_1/z_2) - \frac{1}{12}c_* z_2^{-1}\left(\frac{\partial}{\partial z_1}\right)^3 \delta(z_1/z_2)\end{aligned} \tag{13.44}$$

where

$$c_* = \frac{dl}{l+h}. \tag{13.45}$$

In particular, equating the coefficients of $z_1^{-m-2}z_2^{-n-2}$, *we have*

$$[L_{\mathcal{G}_l}(m), L_{\mathcal{G}_l}(n)] = (m-n)L_{\mathcal{G}_l}(m+n) + \frac{1}{12}(m^3-m)c_*\delta_{m+n,0} \tag{13.46}$$

for $m, n \in \mathbf{Z}$.

Proof From (13.21) and (13.36),

$$\begin{aligned} L_{\mathcal{G}_l}(2)\omega_{\mathcal{G}_l} &= \frac{1}{2(l+h)}\sum_{i=1}^{d}[L_{\mathcal{G}_l}(2), u_i(-1)]u_i(-1)\iota(1) \\ &= \frac{1}{2(l+h)}\sum_{i=1}^{d} u_i(1)u_i(-1)\iota(1) \\ &= \frac{1}{2(l+h)}\sum_{i=1}^{d} l\langle u_i, u_i\rangle\iota(1) \\ &= \frac{dl}{2(l+h)}\iota(1), \end{aligned} \tag{13.47}$$

$$L_{\mathcal{G}_l}(1)\omega_{\mathcal{G}_l} = \frac{1}{2(l+h)}\sum_{i=1}^{d}(u_i(0)u_i(-1)+u_i(-1)u_i(0))\iota(1) = 0, \tag{13.48}$$

$$L_{\mathcal{G}_l}(0)\omega_{\mathcal{G}_l} = \frac{1}{2(l+h)}\sum_{i=1}^{d} 2u_i(-1)^2\iota(1) = 2\omega_{\mathcal{G}_l}. \tag{13.49}$$

That is,

$$\begin{aligned} L_{\mathcal{G}_l}(z)\omega_{\mathcal{G}_l} &= \tfrac{1}{2}c_*\iota(1)z^{-4} + 2\omega_{\mathcal{G}_l}z^{-2} + L_{\mathcal{G}_l}(-1)\omega_{\mathcal{G}_l}z^{-1} \\ &\quad + \text{a nonsingular series in } z. \end{aligned} \tag{13.50}$$

The result now follows from the Jacobi identity and (13.42), just as in (5.83). □

We now have two elements of V_L which generate Virasoro algebras – ω, with central parameter $\dim \mathbf{h} = l \dim \mathcal{H} = l \operatorname{rank} \mathcal{G}$ and $\omega_{\mathcal{G}_l}$, with central parameter $c_* = \frac{dl}{l+h}$. Now it is well known that

$$d = \dim \mathcal{G} = (h+1)\operatorname{rank} \mathcal{G}, \tag{13.51}$$

so that

$$\omega_{\mathcal{G}_l} \neq \omega \quad \text{if} \quad l > 1. \tag{13.52}$$

For $l = 1$ the two central parameters are equal, and in fact we have the following (nontrivial) result:

Proposition 13.10 *For $l = 1$,*

$$\omega_{\mathcal{G}_1} = \omega, \tag{13.53}$$

that is,

$$\frac{1}{2(h+1)} \sum_{i=1}^{d} u_i(-1)^2 \iota(1) = \frac{1}{2} \sum_{i=1}^{\dim \mathcal{H}} v_i(-1)^2 \iota(1), \tag{13.54}$$

where $\{v_1, ..., v_{\dim \mathcal{H}}\}$ is an orthonormal basis of $\mathcal{H}$ with respect to $\langle \cdot, \cdot \rangle$. Equivalently,

$$L_{\mathcal{G}_1}(z) = Y(\omega_{\mathcal{G}_1}, z) = Y(\omega, z) = L(z), \tag{13.55}$$

that is, for all $k \in \mathbf{Z}$,

$$\begin{aligned} \frac{1}{2(h+1)} \sum_{i=1}^{d} \sum_{m \in \mathbf{Z}} :u_i(m) u_i(k-m): &= L_{\mathcal{G}_1}(k) \\ = L(k) = \frac{1}{2} \sum_{i=1}^{\dim \mathcal{H}} \sum_{m \in \mathbf{Z}} &:v_i(m) v_i(k-m):. \end{aligned} \tag{13.56}$$

Proof It suffices to prove (13.53). By (13.36) and (13.39), even for arbitrary $l \geq 1$,

$$[L_{\mathcal{G}_l}(m), x(0)] = 0 = [L(m), x(0)] \tag{13.57}$$

for $x \in \mathcal{G}$ and $m \in \mathbf{Z}$. That is,

$$[\mathcal{G}, L_{\mathcal{G}_l}(z)] = 0 = [\mathcal{G}, L(z)], \tag{13.58}$$

and in particular,

$$x(0)\omega = x(0)L(-2)\iota(1) = L(-2)x(0)\iota(1) = 0 \tag{13.59}$$

and similarly,

$$x(0)\omega_{\mathcal{G}_l} = 0 \tag{13.60}$$

(which we had checked directly in the proof of Proposition 13.5). Thus both $\omega_{\mathcal{G}_l}$ and ω are $\mathcal{G}$-invariant elements of V_L.

Now consider the natural linear map T given by

$$\begin{aligned} T: \mathcal{G} \otimes \mathcal{G} &\rightarrow V_L \\ u \otimes v &\mapsto u(-1)v(-1)\iota(1). \end{aligned} \tag{13.61}$$

As in the proof of Proposition 13.5, this is a $\mathcal{G}$-module map with respect to the natural actions. Denote the image of T by U. Since

$$T((\mathcal{G}\otimes\mathcal{G})^{\mathcal{G}}) = U^{\mathcal{G}}, \tag{13.62}$$

where the superscripts denote the spaces of $\mathcal{G}$-invariants, we have

$$\dim U^{\mathcal{G}} \leq 1. \tag{13.63}$$

On the other hand,

$$\omega_{\mathcal{G}_l} \in U^{\mathcal{G}}, \tag{13.64}$$

and if $l = 1$, we also have

$$\omega \in U^{\mathcal{G}}. \tag{13.65}$$

(Although $\sum_{i=1}^{\dim\mathcal{H}} v_i\otimes v_i \notin (\mathcal{G}\otimes\mathcal{G})^{\mathcal{G}}$, the image under T of this element is invariant.) Thus $\omega_{\mathcal{G}_l}$ and ω are proportional, and using any of the normalizing conditions that we have established, such as the fact that both $L(0)$ and $L_{\mathcal{G}_1}(0)$ fix the element $u(-1)\iota(1)$ for $0 \neq u \in \mathcal{G}$, we see that $L(0) = L_{\mathcal{G}_1}(0)$ and hence that $\omega_{\mathcal{G}_1} = \omega$. □

Remark 13.11: We know from (13.41) that $L(m) = L_{\mathcal{G}_l}(m)$ for $m \geq -1$, and if $l = 1$, this holds for all m, by Proposition 13.10. But for $l > 1$, $L(-2) \neq L_{\mathcal{G}_l}(-2)$, by (13.52). Our discussion shows that in fact

$$\omega \notin L(l,0) \tag{13.66}$$

for $l > 1$, since otherwise, we would have

$$\omega \in U + \mathcal{G}(-2)\iota(1),$$

in the notation of the proof of Proposition 13.9. But the space of $\mathcal{G}$-invariants in this space is just $U^{\mathcal{G}}$, which is spanned by $\omega_{\mathcal{G}_l}$, and ω is not a multiple of $\omega_{\mathcal{G}_l}$.

Now we formulate some of the main conclusions. A vertex operator algebra is *simple* if it is irreducible as a module for itself.

Theorem 13.12 *Let $l > 0$. The structure*

$$L(l,0) = (L(l,0), Y, \iota(1), \omega_{\mathcal{G}_l})$$

is a simple vertex operator algebra of rank $(\dim\mathcal{G})l/(l+h)$. *Its subspace of weight 1, which is identified with $\mathcal{G}$, generates $L(l,0)$ (even without* *the use of the element $\omega_{\mathcal{G}_l}$). The component operators of the vertex operators $Y(x,z)$ for $x \in \mathcal{G}$ together with 1 span a copy of the affine Lie algebra $\hat{\mathcal{G}}$ and $L(l,0)$ is identified with a copy of the level l standard (irreducible) $\hat{\mathcal{G}}$-module whose highest weight space is $\mathcal{G}$-invariant. The component operators of $Y(\omega_{\mathcal{G}_l}, z)$, which generate the Virasoro algebra, are given by the quadratic expressions (13.35), and in fact if W is any module for $L(l,0)$ viewed as a "vertex operator algebra without Virasoro algebra," then $Y(\omega_{\mathcal{G}_l}, z)$ acts on W by means of these same expressions. The vertex operators $Y(x,z)$ are primary fields of weight 1 with respect to $\omega_{\mathcal{G}_l}$ (and this remains true on any W as indicated); equivalently, the Virasoro algebra acts on $\hat{\mathcal{G}}$ according to the bracket relations (13.36). We have $\omega_{\mathcal{G}_l} = \omega$ if and only if $l = 1$.* □

Now the space V_L fails to be an $L(l,0)$-module only in that the eigenspaces of the operator $L_{\mathcal{G}_l}(0)$ do not define the given $\mathbf{Q}$-grading (unless of course $l = 1$). However, V_L is a direct sum of standard (irreducible) $\hat{\mathcal{G}}$-modules of level l, and we now show that each such irreducible $\hat{\mathcal{G}}$-module naturally carries $L(l,0)$-module structure. We know that the level l standard $\hat{\mathcal{G}}$-modules are parametrized by a certain finite set of dominant integral linear forms $\lambda \in \mathcal{H}^*$ (with respect to a fixed choice of positive roots of $\mathcal{G}$ with respect to $\mathcal{H}$) such that the highest weight of the $\hat{\mathcal{G}}$-module, viewed as a linear form on $\mathcal{H} \oplus \mathbf{C}c \subset \hat{\mathcal{G}}$, is given by λ and the correspondence $c \mapsto l$. Let us denote the corresponding standard $\hat{\mathcal{G}}$-module by $L(l,\lambda)$, so that the case $\lambda = 0$ recovers the case $L(l,0)$ that we have been considering. Let W be an irreducible $\hat{\mathcal{G}}$-submodule of V_L and choose λ so that $W \simeq L(l,\lambda)$. By the formula for the action of $L_{\mathcal{G}_l}(0)$ given by (13.35) and the expression (13.20) for the Casimir element Ω of $U(\mathcal{G})$, we see that $L_{\mathcal{G}_l}(0)$ acts on the irreducible $\mathcal{G}$-module generated by the highest weight space of W as the scalar

$$h_{l,\lambda} = \frac{h}{l+h}\Omega_\lambda = \frac{1}{2(l+h)}\langle \lambda, \lambda + 2\rho \rangle \tag{13.67}$$

where Ω_λ is the canonical rational scalar by which Ω acts on the irreducible $\mathcal{G}$-module with highest weight λ and ρ is the half-sum of positive roots. (For instance, the case $\lambda = 0$ gives $h_{l,0} = 0$.) But then the relation

$$[L_{\mathcal{G}_l}(0), x(n)] = -nx(n) \tag{13.68}$$

for $x \in \mathcal{G}$ (see (13.36)) shows that W is $\mathbf{Q}$-graded by the $L_{\mathcal{G}_l}(0)$-

eigenspace decomposition

$$W = \coprod_{n \in h_{l,\lambda} + \mathbf{N}} W_n. \tag{13.69}$$

With respect to this grading, W becomes a module for the vertex operator algebra $L(l, 0)$.

Remark 13.13: It is interesting to note that these considerations hold whether or not the highest weight space of W is homogeneous with respect to the weight grading (by $L(0)$-eigenvalues) of V_L.

Thus we have:

Proposition 13.14 *Every $\hat{\mathcal{G}}$-irreducible subspace of V_L carries the canonical structure of an irreducible $L(l, 0)$-module, where the grading is given by the $L_{\mathcal{G}_l}(0)$-eigenspace decomposition, and the lowest weight is given by (13.67). Equipped with the $\mathbf{Q}$-grading defined by the $L_{\mathcal{G}_l}(0)$-eigenspace decomposition, V_L is the (usually infinite) direct sum of irreducible $L(l, 0)$-modules, and the irreducible $L(l, 0)$-submodules of V_L are exactly the standard $\hat{\mathcal{G}}$-submodules.* □

Remark 13.15: Since the $L_{\mathcal{G}_l}(0)$-eigenspaces of V_L are not in general finite-dimensional, V_L may not strictly speaking be a module for $L(l, 0)$ viewed as a vertex operator algebra, but V_L is a module for $L(l, 0)$ viewed simply as a vertex algebra (where the grading restrictions are ignored).

Next we shall show that every irreducible $L(l, 0)$-module is in fact a standard module. This will be a simple consequence of the following general observation, which follows immediately from formula (13.26) and induction:

Proposition 13.16 *Let V be any vertex operator algebra and let (W, Y_W) be any V-module. Let $v \in V$ be such that the component operators v_n $(n \in \mathbf{Z})$ of $Y_W(v, z)$ all commute with one another, so that $Y_W(v, z)^N$ is well defined on W for $N \in \mathbf{N}$. Then*

$$Y_W((v_{-1})^N \mathbf{1}, z) = Y_W(v, z)^N. \tag{13.70}$$

In particular, if $(v_{-1})^N \mathbf{1} = 0$ for a fixed N, then

$$Y_W(v, z)^N = 0. \tag{13.71}$$ □

Proposition 13.17 *Every $L(l,0)$-module is in a natural way a $\hat{\mathcal{G}}$-module of level l, the $L(l,0)$-invariant subspaces coincide with the $\hat{\mathcal{G}}$-invariant subspaces, and such a module is $L(l,0)$-irreducible if and only if it is a standard $\hat{\mathcal{G}}$-module (of level l).*

Proof We need only prove that an irreducible $L(l,0)$-module (W, Y_W) is a standard $\hat{\mathcal{G}}$-module. Let $U \subset W$ be the $L_{\mathcal{G}_l}(0)$-eigenspace with the lowest eigenvalue. Then U, which is finite-dimensional by assumption, is irreducible under $\mathcal{G}$. Let $w \in U$ be a (nonzero) highest weight vector for $\mathcal{G}$, with respect to a fixed system $\Delta_+ \subset \Delta$ of positive roots of $\mathcal{G}$ with respect to $\mathcal{H}$. Then w is also a highest weight vector for $\hat{\mathcal{G}}$, and W is an irreducible highest-weight $\hat{\mathcal{G}}$-module of level l. Let α_0 be the highest root of $\mathcal{G}$ with respect to $\mathcal{H}$ and Δ_+. It is enough to show that $x_{\alpha_0}(-1)$ acts nilpotently on w, where we use the notation $x(n)$ ($x \in \mathcal{G}$, $n \in \mathbf{Z}$) for the operators defined by $Y_W(x,z) = \sum_{n\in\mathbf{Z}} x(n)z^{-n-1}$. (We already know that $x_{-\alpha}(0)$ acts nilpotently on w for $\alpha \in \Delta_+$ because U is finite-dimensional.)

We show that in fact for $\alpha \in \Delta_+$,

$$x_\alpha(-1)^{l+1}w = 0 \tag{13.72}$$

and

$$x_{-\alpha}(0)^{l+1}w = 0. \tag{13.73}$$

For any $\alpha \in \Delta$,

$$x_\alpha(-1)^{l+1}\iota(1) = 0 \tag{13.74}$$

in the standard level-l module $L(l,0)$ and so

$$Y_W(x_\alpha, z)^{l+1} = 0 \tag{13.75}$$

by Proposition 13.16. But since $x_\alpha(n)w = 0$ for $\alpha \in \Delta_+$ and $n \geq 0$, the coefficient of z^0 in $Y_W(x_\alpha,z)^{l+1}w$ is $x_\alpha(-1)^{l+1}w$, proving (13.72). Analogously, since $x_\alpha(n)w = 0$ for $\alpha \in \Delta$ and $n > 0$, the coefficient of z^{-l-1} in $Y_W(x_\alpha,z)^{l+1}w$ is $x_\alpha(0)^{l+1}w$, which must also vanish, in particular, for $\alpha \in -\Delta_+$, proving (13.73). □

Next we study intertwining operators among modules for the vertex operator algebra $L(l,0)$. By Theorem 12.24 the space $V_L = (V_L, Y, \mathbf{1}, \omega, T, G, F, \Omega)$ (using the notation of that result, applied to the present setting) is an abelian intertwining algebra with $G = L/L_0$. Since $b(0,g) = 0$, $F(0,g_1,g_2) = F(g_1,0,g_2) = 1$ and $\Omega(0,g) = 1$ for

$g, g_1, g_2 \in G$, the operators $Y(v, z_1)$ and $Y(w, z_2)$ satisfy the ordinary Jacobi identity on V_L for $v \in V_L^0 = V_{L_0}$ and $w \in V_L$:

$$\begin{aligned} z_0^{-1}\delta\left(\frac{z_1 - z_2}{z_0}\right)Y(v, z_1)Y(w, z_2) - z_0^{-1}\delta\left(\frac{z_2 - z_1}{-z_0}\right)Y(w, z_2)Y(v, z_1) \\ = z_2^{-1}\delta\left(\frac{z_1 - z_0}{z_2}\right)Y(Y(v, z_0)w, z_2). \end{aligned} \tag{13.76}$$

Also,

$$\frac{d}{dz}Y(w, z) = Y(L(-1)w, z). \tag{13.77}$$

Now V_L is a direct sum of eigenspaces of the operator $L_{\mathcal{G}_l}(0) - L(0)$. We consider the linear operator $z^{L_{\mathcal{G}_l}(0) - L(0)}$ on V_L, defined in the obvious way, so that

$$z^{L_{\mathcal{G}_l}(0) - L(0)}v = z^{L_{\mathcal{G}_l}(0)}z^{-L(0)}v = z^h v \tag{13.78}$$

if $(L_{\mathcal{G}_l}(0) - L(0))v = hv$ for some $h \in \mathbf{Q}$. Using (13.40) and the fact that $L(l, 0)$ is the irreducible $\hat{\mathcal{G}}$-module generated by $\iota(1)$, we have

$$z_1^{L_{\mathcal{G}_l}(0) - L(0)}Y(u, z_2) = Y(u, z_2)z_1^{L_{\mathcal{G}_l}(0) - L(0)} \tag{13.79}$$

as operators on V_L for $u \in L(l, 0)$. For $v \in V_L$ we define a new operator $\mathbf{Y}(v, z)$ on V_L :

$$\mathbf{Y}(v, z) = z^{L_{\mathcal{G}_l}(0) - L(0)}Y(z^{-L_{\mathcal{G}_l}(0) + L(0)}v, z)z^{-L_{\mathcal{G}_l}(0) + L(0)}, \tag{13.80}$$

which gives us a linear map

$$\begin{aligned} \mathbf{Y} : V_L &\to (\mathrm{End}\, V_L)\{z\} \\ v &\mapsto \mathbf{Y}(v, z) = \sum_{n \in \mathbf{Q}} v_n z^{-n-1} \quad (v_n \in \mathrm{End}\, V_L). \end{aligned} \tag{13.81}$$

From (13.41) and (13.79) we see that

$$\mathbf{Y}(v, z) = Y(v, z) \quad \text{for} \quad v \in L(l, 0). \tag{13.82}$$

Recall that V_L is a completely reducible $L(l, 0)$-module. In fact we can and do choose a decomposition

$$V_L = \coprod_{i \in I} W_i \tag{13.83}$$

such that the highest weight vector of each irreducible component W_i is an eigenvector of $L(0) - L_{\mathcal{G}_l}(0)$.

Proposition 13.18 *Consider V_L as a module for $L(l,0)$, viewed as a vertex algebra, as in Proposition 13.14 and Remark 13.15. Then $\mathbf{Y}$ is an intertwining operator of type $\begin{bmatrix} V_L \\ V_L\, V_L \end{bmatrix}$. In particular, for any three irreducible components W_i, W_j and W_k $(i,j,k \in I)$, the projection of $\mathbf{Y}(u,z)v$ to W_i for $u \in W_j$ and $v \in W_k$ is an intertwining operator of type $\begin{bmatrix} W_i \\ W_j\, W_k \end{bmatrix}$. (Here, "intertwining operator" for a vertex algebra or a vertex operator algebra is understood in the sense of (11.5), (11.7), the ordinary Jacobi identity in place of (11.9), and (11.10), as at the beginning of Chapter 12.)*

Proof It suffices to prove the first assertion. The conditions (11.5) and (11.7) are clear. Let w be a highest weight vector of W_j $(j \in I)$. Then $(L_{\mathcal{G}_l}(0) - L(0))w = hw$ for some $h \in \mathbf{Q}$. Using the irreducibility of W_j and (13.40) we see that

$$(L_{\mathcal{G}_l}(0) - L(0))|_{W_j} = h. \tag{13.84}$$

Let $v \in L(l,0) \subset V_L^0$ and $w \in W_j$. Multiplying (13.76) by $z_2^{L_{\mathcal{G}_l}(0)-L(0)-h}$ on the left and by $z_2^{-L_{\mathcal{G}_l}(0)+L(0)}$ on the right, and keeping (13.79), (13.80) and (13.84) in mind we obtain the Jacobi identity

$$\begin{aligned} &z_0^{-1}\delta\left(\frac{z_1-z_2}{z_0}\right)Y(v,z_1)\mathbf{Y}(w,z_2)-z_0^{-1}\delta\left(\frac{z_2-z_1}{-z_0}\right)\mathbf{Y}(w,z_2)Y(v,z_1) \\ &\quad = z_2^{-1}\delta\left(\frac{z_1-z_0}{z_2}\right)\mathbf{Y}(Y(v,z_0)w,z_2). \end{aligned} \tag{13.85}$$

It remains only to prove that $\mathbf{Y}(L_{\mathcal{G}_l}(-1)w,z) = \frac{d}{dz}\mathbf{Y}(w,z)$ for $w \in V_L$. From the Jacobi identity (13.76) and the property (13.77) we have

$$\begin{aligned} [L_{\mathcal{G}_l}(0) - L(0), Y(w,z)] &= Y((L_{\mathcal{G}_l}(0) - L(0))w, z) \\ &\quad + zY((L_{\mathcal{G}_l}(-1) - L(-1))w, z) \end{aligned}$$

as in (6.21), or

$$\begin{aligned} Y(L_{\mathcal{G}_l}(-1)w,z) &= [L_{\mathcal{G}_l}(0) - L(0), Y(w,z)]z^{-1} \\ &\quad - Y((L_{\mathcal{G}_l}(0) - L(0))w,z)z^{-1} + \frac{d}{dz}Y(w,z). \end{aligned} \tag{13.86}$$

Now we take $w \in W_j$. Again multiplying (13.86) by $z^{L_{\mathcal{G}_l}(0)-L(0)-h}$ on the left and by $z^{-L_{\mathcal{G}_l}(0)+L(0)}$ on the right we obtain

$$\begin{aligned}\mathbf{Y}(L_{\mathcal{G}_l}(-1)w,z) &= z^{L_{\mathcal{G}_l}(0)-L(0)}\left(\frac{d}{dz}Y(z^{-L_{\mathcal{G}_l}(0)+L(0)}w,z)\right)z^{-L_{\mathcal{G}_l}(0)+L(0)} \\ &\quad +[L_{\mathcal{G}_l}(0)-L(0),\mathbf{Y}(w,z)]z^{-1} \\ &= \frac{d}{dz}\mathbf{Y}(w,z),\end{aligned}$$

as desired. □

Chapter 14

Z-algebras and parafermion algebras

We continue our discussion in Chapter 13, but we shall relativize it to certain subspaces $\mathbf{h}_*$ of $\mathbf{h}$, following the setting of Chapters 2, 3 and 4. Recall that $\hat{\mathcal{G}}$ is an affine Lie algebra of type $\hat{A}$, $\hat{D}$ or $\hat{E}$ and that l is a positive integer. Restricting our attention to the vacuum space of $L(l,0)$ with respect to a natural Heisenberg subalgebra of $\hat{\mathcal{G}}$, and passing to the quotient spaces of this vacuum space defined by the actions of certain infinite abelian groups, we shall construct canonical generalized vertex operator algebras. Moreover, the corresponding quotient spaces for the vacuum space of any level l standard $\hat{\mathcal{G}}$-module (which can be obtained from a tensor product of basic modules) are modules for these generalized vertex operator algebras. If we make a special choice of the infinite abelian group, the algebra turns out to be simple and the modules turn out to be irreducible. As an illustration, we specialize our construction to the case of the affine Lie algebra $A_1^{(1)}$, and in this case we show in detail the essential equivalence between Z-algebras ([LP1]-[LP3]) and parafermion algebras [ZF1] by realizing the parafermion algebras as canonically modified Z-algebras acting on certain quotient spaces of the vacuum spaces of standard $A_1^{(1)}$-modules defined by the action of an infinite cyclic group (a construction carried out in [LP1]). In particular, we provide a mathematical foundation for "parafermion conformal field theory." In fact, it was our desire to explain and relate Z-algebra theory and parafermion field theory which provided our original motivation for this work. Specifically, it was well known that the Z-algebras satisfy "generalized commutation relations" instead of commutation relations (see [LP1]-[LP3] and [LW2]-[LW5]) and that parafermion algebras satisfy similar generalized commutation relations. In clarifying the equivalence – hinted in [ZF1] – between the notions of Z-algebra and of parafermion algebra, we were led to consider more natural algebras possessing a "generalized Jacobi identity" – what we have called "generalized vertex algebras," defined in Chapters 6 and 9. The generalized commutation relations for both Z-algebras and parafermion algebras follow from special cases of the generalized Jacobi identity by comparing the first few coefficients of z_0.

After recalling the setting of Chapter 13 and choosing a special type of 2-cocycle on the weight lattice of $\mathcal{G}$, we diagonally embed the Cartan subalgebra of $\mathcal{G}$ into the space $\mathbf{h}$, using the setting of Chapters 2-4 with the lattice L taken to be the direct sum of l copies of the weight lattice of $\mathcal{G}$, as in Chapter 13. The image of this diagonal embedding of the Cartan subalgebra is our subspace $\mathbf{h}_*$ of $\mathbf{h}$ with respect to which we relativize the whole structure of Chapter 13. We define B to be a certain infinite abelian subgroup of $\hat{L}_0$ isomorphic to $2Q$, Q being the root lattice of $\mathcal{G}$, and we take the group A of Chapter 4 to be a certain kind of subgroup of B, possibly equal to B. We work out the details of the main constructions of Chapters 2-5 for this setup, and we conclude from Chapters 5 and 6 that we have constructed an example of a generalized vertex operator algebra $\Omega_*^A(0)$ and a family of modules (Theorem 14.2). We point out a parallel, alternative approach (see Remark 14.4) in which we apply the results of Chapter 6 in a different way; for this purpose, we make use of the alternative, more general notion of module defined in Remark 6.15. Using the fact that any $\hat{\mathcal{G}}$-submodule of the space V_L is stable under the group B and related operators (Proposition 14.5), we then construct our basic space $\Omega_{L(l,0)}^A$, which we prove is a "generalized vertex operator algebra without Virasoro algebra" generated by natural elements associated with root vectors of $\mathcal{G}$ (Proposition 14.9). This structure is in fact realized as a substructure of the previously constructed space $\Omega_*^A(0)$, and the result uses a general principle (Proposition 14.8) which precisely describes the subalgebra of a generalized vertex algebra generated by an arbitrary given subset; this general principle, which follows from the Jacobi identity, is a bit more delicate to prove than the corresponding result for (ordinary) vertex operator algebras.

Then we construct the appropriate element of $\Omega_{L(l,0)}^A$ which generates a Virasoro algebra completing the structure of generalized vertex operator algebra on this space. We compute the corresponding weight of the canonical images of the root vectors of $\mathcal{G}$ in this algebra, and we show that these image vectors are lowest weight vectors for the Virasoro algebra and that the corresponding weight space generates the algebra, by analogy with the fact (Chapter 13) that the weight-one subspace of the vertex operator algebra $L(l,0)$, a subspace consisting of lowest weight vectors for the corresponding Virasoro algebra, generates this vertex operator algebra. We next show that for every irreducible $\hat{\mathcal{G}}$-submodule of V_L, the space Ω_M^A constructed by analogy with $\Omega_{L(l,0)}^A$ carries a natural structure of $\Omega_{L(l,0)}^A$-module, and we compute its lowest weight. For the special case in which our abelian group A is the largest possible group B, we show that our algebra $\Omega_{L(l,0)}^B$ is simple and that our modules Ω_M^B are irreducible. In Remark 14.22, we explain the precise correspondence between the structure studied here and the Z-operator structure introduced in [LP1].

In the remainder of this chapter, we work out in detail the important special case in which $\mathcal{G} = \mathbf{sl}(2,\mathbf{C})$. In addition to computing exactly what the general structures and results look like for this case, we take advantage of the special fact that for this case, the group $\hat{L}$ is abelian, allowing us extra flexibility in the basic constructions, especially in the choices of the groups G and A. With our natural new choices for these groups, we must also use the more general alternative notion of module (see Remarks 6.15 and 14.4). The resulting analogues of the main general results are formulated in Theorems 14.24 and 14.25. These results illustrate the flexibility of our axiomatic setup. In addition, as an illustration of the general theory, by specializing the generalized Jacobi identity to very special cases, and introducing notation used in [ZF1], we recover in a precise way the exact structure, operators and formulas of [ZF1], including the equivalence with the corresponding structure introduced in [LP1]-[LP3] and [LW2]-[LW5]; see Remarks 14.26 and 14.34.

Throughout this chapter, we also record in a series of comments the special properties of the degenerate but interesting and nontrivial case $l = 1$, in which the basic generalized vertex operator algebra turns out to have rank 0.

We continue working in the setting of Chapter 13. In particular, we have a simple Lie algebra $\mathcal{G}$ of type A, D or E with nonsingular invariant symmetric bilinear form $\langle\cdot,\cdot\rangle$; $\mathcal{H}$ is a Cartan subalgebra, $\Delta \subset \mathcal{H}$ is the root system, Q and P are the root lattice and weight lattice, respectively; L is the direct sum of l copies of P and $L_0 \subset L$ is the direct sum of l copies of Q, so that L_0 is even and $\langle L, L_0\rangle \subset \mathbf{Z}$; $\hat{L}$ is the central extension of L by $\langle\kappa_q\rangle$ with commutator map c_0, and with $c(\alpha,\beta) = \omega_q^{c_0(\alpha,\beta)}$ for $\alpha,\beta \in L$ and $c(\alpha,\beta) = (-1)^{\langle\alpha,\beta\rangle}$ if $\alpha,\beta \in L_0$; $e\colon L \to \hat{L}$ is a section compatible with (13.1); ϵ_0 is the corresponding 2-cocycle; $\mathbf{h} = \mathbf{C} \otimes_{\mathbf{Z}} L$.

Remark 14.1: It will be convenient to assume that ϵ defined in (13.6) (on P) is biadditive and has the property that $\epsilon(\alpha,\beta) = \pm 1$ if $\alpha,\beta \in Q$. (This can be arranged, with the possible modification of c on L; see Remark 12.18.) Recall from Chapter 13 that the 2-cocycle ϵ_0 on L associated with the section e is the sum of the corresponding 2-cocycles on the P_i (cf. (13.12)). We now have that

$$e_{2\alpha}e_\beta = \epsilon(2\alpha,\beta)e_{2\alpha+\beta} = \epsilon(\alpha,\beta)^2 e_{2\alpha+\beta} = e_{2\alpha+\beta} \tag{14.1}$$

for $\alpha,\beta \in L_0$.

We also work in the setting of Chapters 2-4. Consider the linear map (the "diagonal embedding")

$$\begin{array}{ccc} \mathcal{H} & \xrightarrow{*} & \mathbf{h} \\ \alpha & \mapsto & \alpha_* = \alpha_1 + \cdots + \alpha_l, \end{array} \tag{14.2}$$

where α_i is the element in $\mathbf{C} \otimes_{\mathbf{Z}} P_i \subset \mathbf{h}$ corresponding to $\alpha \in \mathcal{H}$ (cf. (13.12)). Writing Q_* for the image of Q under the map $_*$, we have the group isomorphism

$$_* : Q \to Q_* \subset L_0. \tag{14.3}$$

We also define

$$\mathbf{h}_* = \{\alpha_* | \alpha \in \mathcal{H}\} = \mathcal{H}_*, \tag{14.4}$$

a subspace of $\mathbf{h}$ which we shall take as the space $\mathbf{h}_*$ of Chapters 2-4. Then the orthogonal complement of $\mathbf{h}_*$ in $\mathbf{h}$ is

$$\mathbf{h}_*^\perp = \operatorname{span}\{\alpha_i - \alpha_j | \alpha \in \mathcal{H},\ i,j = 1,\ldots,l\} \tag{14.5}$$

(cf. (2.12)). (Note that if $l = 1$, then $\mathbf{h}_* = \mathbf{h}$.)

Let

$$B = \{e_{\alpha_*} | \alpha \in 2Q\} \subset \hat{L}_0. \tag{14.6}$$

By Remark 14.1, B is actually a subgroup of $\hat{L}_0$, isomorphic to the group $2Q_*$ under the map $\hat{L} \to L$, since

$$e_{\alpha_*}e_{\beta_*} = e_{(\alpha+\beta)_*} \quad \text{for} \quad \alpha,\beta \in 2Q, \tag{14.7}$$

and

$$B \cap \langle \kappa_q \rangle = 1. \tag{14.8}$$

Also, B is clearly central in $\hat{L}_0$, and

$$B \simeq 2Q. \tag{14.9}$$

Let

$$A \subset B \tag{14.10}$$

be an arbitrary (necessarily free abelian) subgroup such that $\bar{A}$ spans $\mathbf{h}_*$, i.e.,

$$\operatorname{rank} A = \dim \mathcal{H} = \operatorname{rank} \mathcal{G}. \tag{14.11}$$

We shall work in the setting of Chapter 4 for the present L, L_0 and A (and $\nu = 1$, $\hat{\nu} = 1$).

In order to apply the results obtained in Chapters 5 and 6, we first observe that the "rational compatibility" assumption (5.30) or (5.31) holds, as does the assumption (5.41), since $\bar{A} \subset 2Q_* \subset 2L_0$. Recalling the notation (2.13), (5.39), (5.43)-(5.44) and (14.2), we have

$$\alpha_i'' = \frac{1}{l}\alpha_* \quad \text{for} \quad \alpha \in \mathcal{H},\ i = 1, ..., l, \tag{14.12}$$

$$L'' = \{\beta'' | \beta \in L\} = \frac{1}{l}P_*,\ L_0'' = \frac{1}{l}Q_*,\ H = L'' \cap 2(L'')^\circ = 2Q_* \tag{14.13}$$

(where P_* is the image of P under the map $_*$) and

$$G = L''/H = \left(\frac{1}{l}P_*\right)/2Q_* \simeq \left(\frac{1}{l}P\right)/2Q. \tag{14.14}$$

(If $l = 1$, then $G = P/2Q$.) In order to determine the smallest positive integer T satisfying (5.34)-(5.35) we first note that

$$\mathbf{Z}\text{−span}\,\langle P, P\rangle = \frac{1}{N}\mathbf{Z}, \tag{14.15}$$

where N is the period of the group P/Q, equal to $n+1$, 4, 2, 3, 2, 1 according as $\mathcal{G}$ is of type A_n $(n \geq 1)$, D_{2n+1} $(n \geq 1)$, D_{2n} $(n \geq 2)$, E_6, E_7, E_8, respectively (cf. for example [Bou]). Thus

$$\mathbf{Z}\text{−span}\,\langle L'', L''\rangle = \mathbf{Z}\text{−span}\,\frac{1}{l}\langle P, P\rangle = \frac{1}{Nl}\mathbf{Z} \tag{14.16}$$

$$(14.17) \qquad \mathbf{Z}\text{–span}\,\langle L, L\rangle = \mathbf{Z}\text{–span}\,\langle P, P\rangle = \frac{1}{N}\mathbf{Z} \subset \frac{1}{Nl}\mathbf{Z}$$

$$(14.18) \quad \langle\gamma',\gamma'\rangle = \langle\gamma,\gamma\rangle - \langle\gamma'',\gamma''\rangle \in 2\mathbf{Z} + \frac{2}{l}\mathbf{Z} = \frac{2}{l}\mathbf{Z} \subset \frac{2}{Nl}\mathbf{Z}$$

for $\gamma \in L_0$, so that

$$(14.19) \qquad T = Nl.$$

The symmetric nondegenerate form (5.45) on G is given by

$$(14.20) \quad (\frac{1}{l}\alpha_* + H, \frac{1}{l}\beta_* + H) = \frac{1}{l}\langle\alpha,\beta\rangle + 2\mathbf{Z} \in \left(\frac{1}{T}\mathbf{Z}\right)/2\mathbf{Z}$$

for $\alpha, \beta \in P$. Recall the structures Ω_*^A, $\Omega_*^A(0)$, $\Omega_*^A(i)$ from (5.59)-(5.63) and (6.39)-(6.40). From Theorems 6.7 and 6.14 together with Remarks 6.5 and 6.6, we have:

Theorem 14.2 *The structure*

$$(14.21) \qquad (\Omega_*^A(0), Y_*^A, \iota(A), \omega \otimes \iota(A), T, G, (\cdot,\cdot))$$

is a generalized vertex operator algebra with rank equal to $(l-1)\mathrm{rank}\,\mathcal{G}$, *and* (Ω_*^A, Y_*^A) *and* $(\Omega_*^A(i), Y_*^A)$ *for* $i \in L/L_0$ *are* $\Omega_*^A(0)$*-modules. The conditions (6.30) and (6.31) relating the* $\frac{1}{T}\mathbf{Z}$*-grading and the* G*-grading of* $\Omega_*^A(0)$ *hold, and the* G*-grading decomposition (6.38) coincides with the character space decomposition of* $\Omega_*^A(0)$ *under the action of* G. □

If $l = 1$, $\Omega_*^A(0) = \mathbf{C}\{Q/\bar{A}\}$, and the structure has rank 0.

Remark 14.3: Note that we cannot invoke the irreducibility result Proposition 5.6 because

$$(14.22) \qquad \bar{A} \subset B = 2Q_*,$$

while

$$(14.23) \qquad L_0 \cap \mathbf{h}_* = Q_*.$$

Remark 14.4: We have applied Theorem 6.7 to the pair of lattices L, L_0. As in the last part of Theorem 6.7 (see (6.42)), we can also apply the theory to the analogous pair L_0, L_0, that is, we can take L_0 to play the role of L in Theorem 6.7, with the group A the same as before. Then by arguing as in (14.15)-(14.19) (and treating $\mathcal{G} \simeq \mathbf{sl}(2,\mathbf{C})$ as an exceptional case), we check easily that

$$(14.24) \qquad T = l,$$

which agrees with the level of the affine Lie algebra representations considered in Chapter 13. Theorem 6.7 then tells us that $\Omega_*^A(0)$ carries a slightly different generalized vertex operator algebra structure than that considered in Theorem 14.2, with $T = l$, for instance. One can then reinterpret the $\Omega_*^A(0)$-modules Ω_*^A and $\Omega_*^A(i)$ in the spirit of Remark 6.15.

We now prepare to develop an analogue of Theorem 13.12. We shall study quotient spaces of certain canonical subspaces of standard modules modulo the action of the group A by embedding them into Ω_*^A and considering them as substructures of Ω_*^A. First we have:

Proposition 14.5 *Any $\hat{\mathcal{G}}$-submodule W of V_L is preserved by the operators e_{α_*} for $\alpha \in \Delta$. Moreover, W is B-stable and in particular, A-stable.*

Proof Recall from (3.38) with $\mathbf{h}_* = 0$ that the vertex operator $Y(a,z)$ for $a \in \hat{L}$ has the expression

$$Y(a,z) = Y(\iota(a),z) = E^-(-\bar{a},z)az^{\bar{a}}E^+(-\bar{a},z) \tag{14.25}$$

where

$$E^{\pm}(h,z) = \exp\left(\sum_{n>0} h(\pm n)z^{\mp n}/(\pm n)\right) \tag{14.26}$$

for $h \in \mathbf{h}$. Thus

$$Y(a,z)\iota(b) = E^-(-\bar{a},z)z^{\langle \bar{a},\bar{b}\rangle}\iota(ab) \tag{14.27}$$

for $b \in \hat{L}$. In particular,

$$\begin{aligned} \iota(a)_{-1}\iota(b) &= \text{coefficient} \\ \text{of } \ z^0 \ \text{ in } \ Y(a,z)\iota(b) &= \begin{cases} 0 & \text{if } \langle \bar{a},\bar{b}\rangle > 0 \\ \iota(ab) & \text{if } \langle \bar{a},\bar{b}\rangle = 0. \end{cases} \end{aligned} \tag{14.28}$$

Hence by (13.16), Proposition 13.1 and (14.2), we have (cf. (13.74))

$$\begin{aligned} \frac{1}{l!}x_\alpha(-1)^l\iota(1) &= \frac{1}{l!}(\iota(e_{\alpha_1})_{-1} + \cdots + \iota(e_{\alpha_l})_{-1})^l\iota(1) \\ &= \iota(e_{\alpha_*}) \in L(l,0) \end{aligned} \tag{14.29}$$

for $\alpha \in \Delta$ (recall the space $L(l,0)$ from (13.19)), where we are using the fact that $e_{\alpha_i}e_{\alpha_j} = e_{\alpha_j}e_{\alpha_i} = e_{\alpha_i+\alpha_j}$ as elements of $\hat{L}$ for $i \neq j$. Since $\iota(e_{\alpha_*}) \in L(l,0)$, we see that the component operators of $Y(e_{\alpha_*}, z)$ preserve W, by Proposition 13.14. Thus the operator

$$e_{\alpha_*} = E^-(\alpha_*, z)Y(e_{\alpha_*}, z)z^{-\alpha_*}E^+(\alpha_*, z) \tag{14.30}$$

preserves W. In particular,

$$e_{2\alpha_*} = \epsilon(\alpha_*, \alpha_*)^{-1}e_{\alpha_*}^2 = \epsilon(\alpha, \alpha)^{-l}e_{\alpha_*}^2 = \pm e_{\alpha_*}^2 \tag{14.31}$$

preserves W. Since B is generated by the elements $e_{2\alpha_*}$ for $\alpha \in \Delta$, W is B-stable and hence A-stable. □

Remark 14.6: From (3.38) with $\mathbf{h}_*$ as in (14.4), the operator e_{α_*} can be expressed as

$$e_{\alpha_*} = Y_*(e_{\alpha_*}, z). \tag{14.32}$$

Next we shall formulate a few definitions. Let M be an $\hat{\mathbf{h}}_*$-submodule of V_L, or equivalently, an $\hat{\mathcal{H}}$-submodule, under the action of $\hat{\mathcal{G}}$ on V_L given by Proposition 13.1. We assume that M is A-stable and that M has an A-stable and $\hat{\mathbf{h}}_*$-stable complement in V_L. As in (4.24), we define

$$W_M = \mathrm{span}\{v - a \cdot v | v \in M, \ a \in A\} \subset M. \tag{14.33}$$

Recall the Heisenberg algebra $(\hat{\mathbf{h}}_*)_{\mathbf{Z}}$ and its vacuum space $\Omega_* \subset V_L$ from Chapter 3. Clearly, $(\hat{\mathbf{h}}_*)_{\mathbf{Z}}$ preserves W_M. Since M has an A-stable complement, we have that

$$M^A = M/W_M \subset V_{L/\bar{A}} \ (= V_L/W_{V_L}) \tag{14.34}$$

(recall (4.23)-(4.24)); the notation M^A is analogous to that in (5.59)-(5.60). Note that the case $M = V_L$ gives $V_L^A = V_{L/\bar{A}}$. Then $(\hat{\mathbf{h}}_*)_{\mathbf{Z}}$ acts naturally on M^A. We set

$$\begin{aligned} \Omega_M^A &= \{v \in M^A | h(n)v = 0 \text{ for } h \in \mathbf{h}_*, \ n > 0\} \\ &= \{v \in M^A | h(n)v = 0 \text{ for } h \in \mathcal{H}, \ n > 0\} \subset \Omega_*^A \end{aligned} \tag{14.35}$$

(recall (4.27) and (5.59)), where for $h \in \mathcal{H}$, $h(n)$ is the operator $\pi(h \otimes t^n) = h_*(n)$ defined in Chapter 13. Then Ω_M^A is also the obvious quotient of $\Omega_* \cap M$:

$$\Omega_M^A = (\Omega_* \cap M)/(\Omega_* \cap W_M). \tag{14.36}$$

Remark 14.7: By Proposition 14.5 and the fact that V_L is a completely reducible $\hat{\mathcal{G}}$-module, any $\hat{\mathcal{G}}$-submodule M of V_L satisfies the conditions above. Taking $M = L(l,0)$, we observe that for $\alpha \in \Delta$,

$$i(x_\alpha) \in \Omega_* \cap L(l,0) \tag{14.37}$$

(recall (13.16)), so that

$$i(x_\alpha) + W_{L(l,0)} \in \Omega^A_{L(l,0)}, \tag{14.38}$$

in view of (14.36).

Next we shall show that $\Omega^A_{L(l,0)}$ is the "generalized vertex operator subalgebra without Virasoro algebra" of $\Omega^A_*(0)$ (in the obvious sense) generated by $i(x_\alpha) + W_{L(l,0)}$ for $\alpha \in \Delta$, where we use the same T, G and $(\cdot,\cdot)$ as for $\Omega^A_*(0)$ (recall Theorem 14.2). For this, we shall use the following general principle (cf. [FHL], (2.4.6)):

Proposition 14.8 *Let $V = (V, Y, \mathbf{1}, \omega, T, G, (\cdot,\cdot), c(\cdot,\cdot))$ be a generalized vertex algebra as defined in Chapter 9. The subalgebra $\langle S\rangle$ generated by a subset S of V, defined as the smallest subalgebra containing S, is given by*

$$\begin{aligned}\langle S\rangle = \operatorname{span}\{v^{(1)}_{n_1}\cdots v^{(j)}_{n_j}\cdot v^{(j+1)} | &\text{each } v^{(i)} \in \{\text{homogeneous}\\ &\text{components of elements of } S,\\ &\text{with respect to the double grading}\} \cup \{\mathbf{1},\omega\}\},\end{aligned} \tag{14.39}$$

and the subalgebra without Virasoro algebra generated by S is given similarly, but with ω omitted from the right-hand side of (14.39).

Proof It suffices to show that for u, v and w (doubly) homogeneous elements of V and for $m, n \in \mathbf{Q}$, the element $(u_m v)_n w$ of V can be expressed as a finite linear combination of elements of the form $u_p v_q w$ and $v_p u_q w$ ($p, q \in \mathbf{Q}$). For this, we use the generalized associator relation (7.11), which remains valid for generalized vertex algebras, with the insertion of the factor $c(g,h)$, as is indicated in Chapter 9. Assuming that $w \in V^k$ ($k \in G$), we see that the first term on the left-hand side of (7.11) gives

$$Y(Y(u,z_0)v,z_2)w(z_2+z_0)^{(g,k)}z_0^{(g,h)}z_2^{(h,k)}. \tag{14.40}$$

We consider the expansion

$$(14.41) \qquad Y(u,z_0)v = \sum_{m\in\frac{1}{T}\mathbf{Z},\, m\le m_0} u_m v z_0^{-m-1}$$

and we assume that

$$(14.42) \qquad u_{m_0}v \neq 0.$$

(If $Y(u,z_0)v = 0$, there is nothing to prove.) Equating the coefficients of

$$z_0^{-m_0-1+(g,h)} z_2^{-n-1+(g,k)+(h,k)} \quad (n \in \mathbf{Q})$$

in (7.11) shows that $(u_{m_0}v)_n w$ can be expressed as desired. We now proceed by induction on $s = 1, 2, \ldots$ to show that $(u_{m_0-s/T}v)_n w$ $(n \in \mathbf{Q})$ can be so expressed. □

Proposition 14.9 *The space $\Omega^A_{L(l,0)}$ is a "generalized vertex operator subalgebra without Virasoro algebra" of $\Omega^A_*(0)$, and in fact is precisely the subalgebra generated by the elements $i(x_\alpha) + W_{L(l,0)}$ for $\alpha \in \Delta$.*

Proof Set

$$(14.43) \quad \Omega_{L(l,0)} = \{v \in L(l,0) | h(n)v = 0 \text{ for } h \in \mathbf{h}_*, n > 0\} = \Omega_* \cap L(l,0)$$

and

$$U = \text{span}\left\{v^{(1)}_{n_1} \cdots v^{(j)}_{n_j} \cdot \iota(1) | \text{each } v^{(i)} \in \{i(x_\alpha) | \alpha \in \Delta\} \cup \{\iota(1)\}\right\},$$

where $v^{(i)}_{n_i}$ is understood as a component operator of $Y_*(v^{(i)}, z)$ (see (3.41)). We first prove that

$$(14.44) \qquad \Omega_{L(l,0)} = U.$$

Comparing the definition (3.38) of the relative vertex operators $Y_*(a,z)$ for $a \in \hat{L}$ with the definition (14.25) and recalling (13.16), (14.2) and (14.26), we see that

$$(14.45) \quad Y_*(i(x_\alpha), z) = E^-(\frac{1}{l}\alpha_*, z) Y(i(x_\alpha), z) z^{-\alpha_*/l} E^+(\frac{1}{l}\alpha_*, z),$$

since the projections of $\alpha_1, \ldots, \alpha_l$ to $\mathbf{h}_*$ are all equal to α_*/l. Thus the component operators of $Y_*(i(x_\alpha), z)$ preserve $L(l,0)$, in view of Proposition 13.1. Recall from (3.46) the basic fact that the Heisenberg algebra

$(\hat{\mathbf{h}}_*)_{\mathbf{Z}}$ commutes with the relative vertex operators, including the operators (14.45). Since (because of this commutativity) the component operators of $Y_*(\iota(x_\alpha), z)$ also preserve Ω_* (recall (3.49)), we conclude that $\Omega_{L(l,0)}$ is stable under the component operators of $Y_*(\iota(x_\alpha), z)$. In particular, U is a subspace of $\Omega_{L(l,0)}$. On the other hand, from (14.45) we have the reverse relation

$$(14.46)\qquad Y(\iota(x_\alpha), z) = E^-(-\frac{1}{l}\alpha_*, z)Y_*(\iota(x_\alpha), z)z^{\alpha_*/l}E^+(-\frac{1}{l}\alpha_*, z).$$

It follows that

$$(14.47)\qquad U((\hat{\mathbf{h}}_*)_{\mathbf{Z}})U = U(\hat{\mathbf{h}}_*^-) \otimes U \subset U(\hat{\mathbf{h}}_*^-) \otimes \Omega_{L(l,0)} = L(l,0)$$

(recall (3.35)-(3.36)), which is stable under $(\hat{\mathbf{h}}_*)_{\mathbf{Z}}$, under the component operators of each $Y_*(\iota(x_\alpha), z)$ and under the operator α_*, is a $\hat{\mathcal{G}}$-submodule of $L(l,0)$. By the irreducibility of $L(l,0)$, $U((\hat{\mathbf{h}}_*)_{\mathbf{Z}})U = L(l,0)$ and consequently $\Omega_{L(l,0)} = U$.

By (14.35), (14.36), Remark 14.7 and (14.43),

$$(14.48)\qquad \Omega^A_{L(l,0)} = \Omega_{L(l,0)}/(\Omega_* \cap W_{L(l,0)}) \subset \Omega^A_*(0),$$

and from (14.44) and (4.39),

$$(14.49)\qquad \begin{aligned} \Omega^A_{L(l,0)} = \operatorname{span}\Big\{ & v^{(1)}_{n_1} \cdots v^{(j)}_{n_j} \cdot (\iota(1) + W_{L(l,0)}) | \text{each } v^{(i)} \\ & \in \{\iota(x_\alpha) + W_{L(l,0)} | \alpha \in \Delta\} \cup \{\iota(1) + W_{L(l,0)}\}\Big\}. \end{aligned}$$

Since the elements $\iota(x_\alpha) + W_{L(l,0)}$ for $\alpha \in \Delta$ are homogeneous with respect to the G-grading (recall (5.46), (14.12) and (14.14)) and with respect to the $\mathbf{Q}$-grading (recall (4.15) and (5.38); the details are worked out in Proposition 14.15 below), the result follows from Proposition 14.8. □

We shall next find a Virasoro element in $\Omega^A_{L(l,0)}$ so that $\Omega^A_{L(l,0)}$ will be a generalized vertex operator algebra.

The usual Virasoro element associated with the affine algebra $\hat{\mathbf{h}}_*$ is

$$(14.50)\qquad \omega_{\mathcal{H}_l} = \frac{1}{2}\sum_{i=1}^{r} \gamma_i(-1)^2\iota(1) \in L(l,0)$$

where $r = \dim \mathbf{h}_* = \dim \mathcal{H}$ and $\{\gamma_1, ..., \gamma_r\}$ is an orthonormal basis of $\mathbf{h}_*$ (cf. (5.65)). Then for $h \in \mathbf{h}_*$ and $m, n \in \mathbf{Z}$,

$$(14.51)\qquad [L_{\mathcal{H}_l}(m), h(n)] = -nh(m+n)$$

(cf. (5.70)), where

$$Y(\omega_{\mathcal{H}_l}, z) = L_{\mathcal{H}_l}(z) = \sum_{m \in \mathbf{Z}} L_{\mathcal{H}_l}(m) z^{-m-2}. \tag{14.52}$$

Recall (13.21) and set

$$\omega_{\mathcal{G}_l, \mathcal{H}_l} = \omega_{\mathcal{G}_l} - \omega_{\mathcal{H}_l} \in L(l, 0). \tag{14.53}$$

Then from (13.34), (13.36) and (14.51),

$$[h(n), L_{\mathcal{G}_l}(m) - L_{\mathcal{H}_l}(m)] = 0 \quad \text{on} \quad V_L \tag{14.54}$$

(cf. (13.40)). (This is another case of a result in the "coset construction," this time, for the pair $(\mathcal{G}, \mathcal{H})$; recall Remark 13.7. Below we shall also show that $\omega_{\mathcal{G}_l, \mathcal{H}_l}$ generates a Virasoro algebra; cf. [GKO].) Thus

$$h(n)\omega_{\mathcal{G}_l, \mathcal{H}_l} = h(n)(L_{\mathcal{G}_l}(-2) - L_{\mathcal{H}_l}(-2))\iota(1)) = 0 \tag{14.55}$$

for $h \in \mathbf{h}_*$ and $n \geq 0$, and so

$$\omega_{\mathcal{G}_l, \mathcal{H}_l} \in \Omega_* \tag{14.56}$$

(recall (3.29)). By (3.33), we can write

$$\omega_{\mathcal{G}_l, \mathcal{H}_l} = \sum_{i=1}^{n} u_i \otimes \iota(a_i) \tag{14.57}$$

for some $n > 0$, $0 \neq u_i \in S((\hat{\mathbf{h}}_*^{\perp})^-)$ and $a_i \in \hat{L}$ such that the $\iota(a_i)$ are linearly independent. Then

$$0 = h(0)\omega_{\mathcal{G}_l, \mathcal{H}_l} = \sum_{i=1}^{n} \langle h, \bar{a}_i \rangle u_i \otimes \iota(a_i) \tag{14.58}$$

for $h \in \mathbf{h}_*$ and so $\langle h, \bar{a}_i \rangle = 0$. Thus the projection $\bar{a}_i''$ of $\bar{a}_i$ to $\mathbf{h}_*$ is 0 for $i = 1, ..., n$, or equivalently,

$$\bar{a}_i' = \bar{a}_i \tag{14.59}$$

(see (2.13)). By the definitions (3.38) and (3.40) of vertex operators, we immediately have:

Proposition 14.10 *On* V_L,

$$Y_*(\omega_{\mathcal{G}_l}, z) = Y_*(\omega_{\mathcal{G}_l, \mathcal{H}_l}, z) = Y(\omega_{\mathcal{G}_l, \mathcal{H}_l}, z) = Y(\omega_{\mathcal{G}_l}, z) - Y(\omega_{\mathcal{H}_l}, z). \quad \square \tag{14.60}$$

Note that if $l = 1$, $\omega_{\mathcal{G}_l,\mathcal{H}_l} = 0$ (recall Proposition 13.10).

Recall from (5.65) that the Virasoro element associated with the affine algebra $\hat{\mathbf{h}}_*^\perp$ is

$$\omega_{\mathbf{h}_*^\perp} = \frac{1}{2}\sum_{i=1}^{s} \beta_i(-1)^2\iota(1), \tag{14.61}$$

where $\{\beta_1, ..., \beta_s\}$ is an orthonormal basis of $\mathbf{h}_*^\perp$. Also recall from Chapter 13 that ω is the Virasoro element of the vertex operator algebra V_{L_0}. Then

$$\omega = \omega_{\mathcal{H}_l} + \omega_{\mathbf{h}_*^\perp} \tag{14.62}$$

and

$$L(z) = Y(\omega, z) = L_{\mathcal{H}_l}(z) + L_{\mathbf{h}_*^\perp}(z), \tag{14.63}$$

or equivalently,

$$L(m) = L_{\mathcal{H}_l}(m) + L_{\mathbf{h}_*^\perp}(m) \quad \text{for} \quad m \in \mathbf{Z}, \tag{14.64}$$

where

$$L_{\mathbf{h}_*^\perp}(z) = Y(\omega_{\mathbf{h}_*^\perp}, z) = Y_*(\omega_{\mathbf{h}_*^\perp}, z) = \sum_{m\in\mathbf{Z}} L_{\mathbf{h}_*^\perp}(m)z^{-m-2}. \tag{14.65}$$

We set

$$L_{\mathcal{G}_l,\mathcal{H}_l}(z) = Y_*(\omega_{\mathcal{G}_l,\mathcal{H}_l}, z) = \sum_{m\in\mathbf{Z}} L_{\mathcal{G}_l,\mathcal{H}_l}(m)z^{-m-2}. \tag{14.66}$$

Proposition 14.11 *For $m \geq -1$,*

$$L_{\mathbf{h}_*^\perp}(m) = L_{\mathcal{G}_l,\mathcal{H}_l}(m) \tag{14.67}$$

on $L(l, 0)$. In particular, $L_{\mathcal{G}_l,\mathcal{H}_l}(0)$ defines the giving grading of $\Omega^A_{L(l,0)}$, and for $v \in L(l, 0)$,

$$Y_*(L_{\mathcal{G}_l,\mathcal{H}_l}(-1)v, z) = \frac{d}{dz}Y_*(v, z) \tag{14.68}$$

on V_L.

Proof The second assertion is a direct consequence of the first assertion and the $L_{\mathbf{h}_*^\perp}(0)$-grading property (see Theorem 6.7) and $L_{\mathbf{h}_*^\perp}(-1)$-property (5.71). By Proposition 13.8, $L(m)$ and $L_{\mathcal{G}_l}(m)$ agree on $L(l,0)$ for $m \geq -1$. Thus $L_{\mathbf{h}_*^\perp}(m) = L(m) - L_{\mathcal{H}_l}(m)$ and $L_{\mathcal{G}_l,\mathcal{H}_l}(m) = L_{\mathcal{G}_l}(m) - L_{\mathcal{H}_l}(m)$ agree on $L(l,0)$. □

From (14.53) and (14.56), $\omega_{\mathcal{G}_l,\mathcal{H}_l} \in \Omega_{L(l,0)}$, so that

$$\omega_{\mathcal{G}_l,\mathcal{H}_l} + W_{L(l,0)} \in \Omega^A_{L(l,0)}, \tag{14.69}$$

and as we shall see below, this vector will be the appropriate Virasoro element for $\Omega^A_{L(l,0)}$. Using the notation (4.39), write

$$L^A_{\mathcal{G}_l,\mathcal{H}_l}(z) = Y^A_*(\omega_{\mathcal{G}_l,\mathcal{H}_l} + W_{L(l,0)}, z) = \sum_{k\in\mathbf{Z}} L^A_{\mathcal{G}_l,\mathcal{H}_l}(k)z^{-k-2}. \tag{14.70}$$

The following is an immediate consequence of Proposition 14.11:

Proposition 14.12 *On $V_L^A = V_{L/\bar{A}}$,*

$$Y^A_*(L^A_{\mathcal{G}_l,\mathcal{H}_l}(-1)v, z) = \frac{d}{dz}Y^A_*(v,z) \tag{14.71}$$

for $v \in \Omega^A_{L(l,0)}$. □

Proposition 14.13 *On V_L,*

$$\begin{aligned} [L_{\mathcal{G}_l,\mathcal{H}_l}(z_1), L_{\mathcal{G}_l,\mathcal{H}_l}(z_2)] &= z_2^{-1}\left(\frac{d}{dz_2}L_{\mathcal{G}_l,\mathcal{H}_l}(z_2)\right)\delta(z_1/z_2) \\ &-2z_2^{-1}L_{\mathcal{G}_l,\mathcal{H}_l}(z_2)\frac{\partial}{\partial z_1}\delta(z_1/z_2) - \frac{1}{12}c_{\mathcal{G}_l,\mathcal{H}_l}z_2^{-1}\left(\frac{\partial}{\partial z_1}\right)^3\delta(z_1/z_2) \end{aligned} \tag{14.72}$$

where

$$c_{\mathcal{G}_l,\mathcal{H}_l} = \frac{dl}{l+h} - r, \tag{14.73}$$

$r = \operatorname{rank}\mathcal{G}$, $d = \dim\mathcal{G}$ and h is the Coxeter number of $\mathcal{G}$, as in Chapter 13. In particular, equating the coefficients of $z_1^{-m-2}z_2^{-n-2}$, we have

$$\begin{aligned} [L_{\mathcal{G}_l,\mathcal{H}_l}(m), L_{\mathcal{G}_l,\mathcal{H}_l}(n)] &= (m-n)L_{\mathcal{G}_l,\mathcal{H}_l}(m+n) \\ &+\frac{1}{12}(m^3-m)c_{\mathcal{G}_l,\mathcal{H}_l}\delta_{m+n,0} \end{aligned} \tag{14.74}$$

for $m, n \in \mathbf{Z}$.

Proof First, we have that

$$[L_{\mathcal{G}_l}(z_1), L_{\mathcal{G}_l}(z_2)] = z_2^{-1}\left(\frac{d}{dz_2}L_{\mathcal{G}_l}(z_2)\right)\delta(z_1/z_2)$$

$$-2z_2^{-1}L_{\mathcal{G}_l}(z_2)\frac{\partial}{\partial z_1}\delta(z_1/z_2) - \frac{1}{12}c_* z_2^{-1}\left(\frac{\partial}{\partial z_1}\right)^3\delta(z_1/z_2)$$

and that

$$[L_{\mathcal{H}_l}(z_1), L_{\mathcal{H}_l}(z_2)] = z_2^{-1}\left(\frac{d}{dz_2}L_{\mathcal{H}_l}(z_2)\right)\delta(z_1/z_2)$$

$$-2z_2^{-1}L_{\mathcal{H}_l}(z_2)\frac{\partial}{\partial z_1}\delta(z_1/z_2) - \frac{1}{12}r z_2^{-1}\left(\frac{\partial}{\partial z_1}\right)^3\delta(z_1/z_2),$$

by Proposition 13.9 and (5.83). By (14.54), (14.60) and the definition of the vertex operator $L_{\mathcal{H}_l}(z)$, we find that

$$[L_{\mathcal{G}_l,\mathcal{H}_l}(z_1), L_{\mathcal{H}_l}(z_2)] = [L_{\mathcal{H}_l}(z_1), L_{\mathcal{G}_l,\mathcal{H}_l}(z_2)] = 0. \tag{14.75}$$

Thus

$$\begin{aligned}
&[L_{\mathcal{G}_l,\mathcal{H}_l}(z_1), L_{\mathcal{G}_l,\mathcal{H}_l}(z_2)] = [L_{\mathcal{G}_l,\mathcal{H}_l}(z_1), L_{\mathcal{G}_l}(z_2)]\\
&= [L_{\mathcal{G}_l}(z_1), L_{\mathcal{G}_l}(z_2)] - [L_{\mathcal{H}_l}(z_1), L_{\mathcal{G}_l}(z_2)]\\
&= [L_{\mathcal{G}_l}(z_1), L_{\mathcal{G}_l}(z_2)] - [L_{\mathcal{H}_l}(z_1), L_{\mathcal{H}_l}(z_2)]\\
&= z_2^{-1}\left(\frac{d}{dz_2}L_{\mathcal{G}_l,\mathcal{H}_l}(z_2)\right)\delta(z_1/z_2) - 2z_2^{-1}L_{\mathcal{G}_l,\mathcal{H}_l}(z_2)\frac{\partial}{\partial z_1}\delta(z_1/z_2)\\
&\quad -\frac{1}{12}c_{\mathcal{G}_l,\mathcal{H}_l}z_2^{-1}\left(\frac{\partial}{\partial z_1}\right)^3\delta(z_1/z_2),
\end{aligned} \tag{14.76}$$

as desired. □

Thus we have:

Proposition 14.14 *On* V_L^A,

$$\begin{aligned}
&[L^A_{\mathcal{G}_l,\mathcal{H}_l}(z_1), L^A_{\mathcal{G}_l,\mathcal{H}_l}(z_2)] = z_2^{-1}\left(\frac{d}{dz_2}L^A_{\mathcal{G}_l,\mathcal{H}_l}(z_2)\right)\delta(z_1/z_2)\\
&-2z_2^{-1}L^A_{\mathcal{G}_l,\mathcal{H}_l}(z_2)\frac{\partial}{\partial z_1}\delta(z_1/z_2) - \frac{1}{12}c_{\mathcal{G}_l,\mathcal{H}_l}z_2^{-1}\left(\frac{\partial}{\partial z_1}\right)^3\delta(z_1/z_2).
\end{aligned} \tag{14.77}$$

In particular,

$$[L^A_{\mathcal{G}_l,\mathcal{H}_l}(m), L^A_{\mathcal{G}_l,\mathcal{H}_l}(n)] = (m-n)L^A_{\mathcal{G}_l,\mathcal{H}_l}(m+n) + \frac{1}{12}(m^3-m)c_{\mathcal{G}_l,\mathcal{H}_l}\delta_{m+n,0} \tag{14.78}$$

for $m, n \in \mathbf{Z}$. □

We also need:

Proposition 14.15 *For* $\alpha \in \Delta$, *the vector* $i(x_\alpha) + W_{L(l,0)}$ *is an eigenvector for the operator* $L^A_{\mathcal{G}_l,\mathcal{H}_l}(0)$ *with eigenvalue* $1-\frac{1}{l}$, *that is,*

$$L^A_{\mathcal{G}_l,\mathcal{H}_l}(0)(i(x_\alpha) + W_{L(l,0)}) = (1-\frac{1}{l})(i(x_\alpha) + W_{L(l,0)}). \tag{14.79}$$

Moreover, $1-\frac{1}{l}$ *is also the degree of* $i(x_\alpha) + W_{L(l,0)}$ *with respect to the grading (5.38). This vector is a lowest weight vector for the Virasoro algebra given in Proposition 14.14.*

Proof For the first assertion, it suffices by (14.60) to prove that

$$L_{\mathcal{G}_l,\mathcal{H}_l}(0)i(x_\alpha) = (L_{\mathcal{G}_l}(0) - L_{\mathcal{H}_l}(0))i(x_\alpha) = (1-\frac{1}{l})i(x_\alpha). \tag{14.80}$$

But

$$L_{\mathcal{G}_l}(0)i(x_\alpha) = i(x_\alpha)$$

(see Chapter 13), and by (13.16) and (14.12),

$$\begin{aligned} L_{\mathcal{H}_l}(0)i(x_\alpha) &= L_{\mathcal{H}_l}(0)(\iota(e_{\alpha_1}) + \cdots + \iota(e_{\alpha_l})) \\ &= \frac{1}{2}\langle\alpha_1'', \alpha_1''\rangle\iota(e_{\alpha_1}) + \cdots + \frac{1}{2}\langle\alpha_l'', \alpha_l''\rangle\iota(e_{\alpha_l}) \\ &= \frac{1}{2}\langle\frac{1}{l}\alpha_*, \frac{1}{l}\alpha_*\rangle i(x_\alpha) \\ &= \frac{1}{l}i(x_\alpha), \end{aligned}$$

as desired. The same computation proves the second assertion as well (recall (4.15)). (The second assertion also follows from Proposition 14.11.) The last assertion follows immediately from weight considerations, for instance. □

We have the following analogue of Theorem 13.12:

Theorem 14.16 *The space*

$$(\Omega^A_{L(l,0)}, Y^A_*, \iota(A), \omega_{\mathcal{G}_l,\mathcal{H}_l} \otimes \iota(A), T, G, (\cdot,\cdot))$$

is a generalized vertex operator algebra with rank equal to $\frac{dl}{l+h} - r$, *generated by the* $L^A_{\mathcal{G}_l,\mathcal{H}_l}(0)$*-eigenspace of* $\Omega^A_{L(l,0)}$ *with eigenvalue* $1-\frac{1}{l}$. *The vertex operators* $Y^A_*(i(x_\alpha)+W_{L(l,0)}, z)$ *for* $\alpha \in \Delta$ *are primary fields of weight* $1-\frac{1}{l}$. □

Note that if $l=1$, the rank is 0 and $\Omega^A_{L(l,0)} = \Omega^A_*(0) = \mathbf{C}\{Q/\bar{A}\}$.

We next show that for any irreducible $\hat{\mathcal{G}}$-submodule M of V_L, the subspace Ω^A_M (recall (14.35) and Remark 14.7) of Ω^A_* naturally carries $\Omega^A_{L(l,0)}$-module structure. We shall take $M \simeq L(l,\lambda)$, with $\lambda \in \mathcal{H}^*$ (recall the discussion preceding (13.67)). From Theorem 14.2, (Ω^A_*, Y^A_*) is an $\Omega^A_*(0)$-module. In particular, (Ω^A_*, Y^A_*) is a module for $\Omega^A_{L(l,0)}$ viewed as a "generalized vertex operator algebra without Virasoro algebra." What we need to prove is that Ω^A_M is G-graded, that Ω^A_M is also $\mathbf{Q}$-graded by the $L^A_{\mathcal{G}_l,\mathcal{H}_l}(0)$-eigenspace decomposition, that each eigenspace is finite-dimensional and that the component operators of $Y^A_*(v,z)$ for $v \in \Omega^A_{L(l,0)}$ preserve Ω^A_M.

First note that

$$\Omega_* = \coprod_{\mu\in\frac{1}{l}P} \Omega^{\mu_*}_* \tag{14.81}$$

where

$$\begin{aligned}\Omega^{\mu_*}_* &= \{v \in \Omega_* | h(0)v = \langle h_*, \mu_*\rangle v \ \text{ for } \ h \in \mathcal{H}\} \\ &= \sum_{a\in\hat{L},\ \bar{a}''=\mu_*} S((\hat{\mathfrak{h}}^\perp_*)^-) \otimes \iota(a)\end{aligned} \tag{14.82}$$

(recall (3.33) and (14.12)-(14.13)). Since $\Omega_M = M \cap \Omega_*$ is stable under the operators $h(0)$ for $h \in \mathcal{H}$, we see that

$$\Omega_M = \coprod_{\mu\in\frac{1}{l}P} \Omega^{\mu_*}_M \tag{14.83}$$

where

$$\Omega^{\mu_*}_M = \Omega_M \cap \Omega^{\mu_*}_*. \tag{14.84}$$

Thus Ω^A_M is G-graded, with

$$(\Omega^A_M)^g = \sum_{\mu\in\frac{1}{l}P,\ \mu_*+2Q_*=g} (\Omega^{\mu_*}_M + W_M)/W_M \tag{14.85}$$

for $g \in G$ (recall (5.46) and (14.13)-(14.14)). Moreover, for any highest weight vector w_0 of M, its image $w_0 + W_M$ in Ω_M^A is G-homogeneous.

The same argument as in the proof of Proposition 14.9 shows that

$$\begin{aligned}\Omega_M^A &= \operatorname{span}\Big\{v_{n_1}^{(1)} \cdots v_{n_j}^{(j)} \cdot (w_0 + W_M) | \text{each } v^{(i)} \\ &\in \{i(x_\alpha) + W_{L(l,0)} | \alpha \in \Delta\} \cup \{\iota(1) + W_{L(l,0)}\}\Big\}\end{aligned} \tag{14.86}$$

for any highest weight vector w_0 of M, just as in (14.49). In particular, Ω_M^A is stable under the component operators of $Y_*^A(i(x_\alpha) + W_{L(l,0)}, z)$ for $\alpha \in \Delta$. Clearly, w_0 is an eigenvector for both $L_{\mathcal{G}_l}(0)$ and $L_{\mathcal{H}_l}(0)$, so that $w_0 + W_M$ is an eigenvector for $L_{\mathcal{G}_l,\mathcal{H}_l}^A(0)$ and hence is doubly homogeneous. In fact, recalling that we have taken $M \simeq L(l, \lambda)$, we see that the $L_{\mathcal{G}_l,\mathcal{H}_l}^A(0)$-eigenvalue is $h_{l,\lambda} - \frac{1}{2l}\langle\lambda, \lambda\rangle$, by using the fact that $w_0 \in \Omega_*^{\frac{1}{l}\lambda_*}$, in the notation of (14.82), and by computing as in the proof of Proposition 14.15 (recall the notation $h_{l,\lambda}$ from (13.67)). By (14.71), (14.79) and the Jacobi identity,

$$[L_{\mathcal{G}_l,\mathcal{H}_l}^A(0), v_n] = (-n - \frac{1}{l})v_n \tag{14.87}$$

for $v = i(x_\alpha) + W_{L(l,0)}$ ($\alpha \in \Delta$) and $n \in \mathbf{Q}$, as in Remark 6.11, for instance. Thus from (14.86), Ω_M^A is $\mathbf{Q}$-graded by the $L_{\mathcal{G}_l,\mathcal{H}_l}^A(0)$-eigenspace decomposition. Clearly, $w_0 + W_M \in \Omega_M^A$ is a sum of $L_{\mathbf{h}_*^\perp}(0)$-eigenvectors (recall the relative vertex operator $L_{\mathbf{h}_*^\perp}(z)$ from (14.65)). Using Proposition 14.11 and the Jacobi identity we find that

$$[L_{\mathbf{h}_*^\perp}(0), v_n] = [L_{\mathcal{G}_l,\mathcal{H}_l}(0), v_n] \tag{14.88}$$

for $v = i(x_\alpha)$ ($\alpha \in \Delta$) and $n \in \mathbf{Q}$. Thus by (14.86), any $L_{\mathcal{G}_l,\mathcal{H}_l}^A(0)$-eigenspace of Ω_M^A is contained in a finite sum of $L_{\mathbf{h}_*^\perp}(0)$-eigenspaces and hence is finite-dimensional (by Theorem 14.2). Since $\Omega_{L(l,0)}^A$ is generated by the elements $i(x_\alpha) + W_{L(l,0)}$ (Proposition 14.9), the argument in the proof of Proposition 14.8, with w replaced by the (doubly homogeneous) elements on the right-hand side of (14.86), shows that Ω_M^A is stable under the component operators of $Y_*^A(v, z)$ for $v \in \Omega_{L(l,0)}^A$. Thus we have the following analogue of Proposition 13.14, where we use the notation (13.67):

Proposition 14.17 *For any irreducible $\hat{\mathcal{G}}$-submodule $M \simeq L(l, \lambda)$ ($\lambda \in \mathcal{H}^*$) of V_L, the subspace Ω_M^A of Ω_*^A carries the structure of an*

$\Omega^A_{L(l,0)}$-module, where the grading is given by the $L^A_{\mathcal{G}_l,\mathcal{H}_l}(0)$-eigenspace decomposition, and the lowest weight is given by

$$(14.89)\quad h^*_{l,\lambda} = h_{l,\lambda} - \frac{1}{2l}\langle\lambda,\lambda\rangle = \frac{1}{2(l+h)}\langle\lambda,\lambda+2\rho\rangle - \frac{1}{2l}\langle\lambda,\lambda\rangle. \ \square$$

For the case $l = 1$, $h^*_{l,\lambda} = 0$, which can be checked directly from the fact that $2\langle\lambda,\rho\rangle = h\langle\lambda,\lambda\rangle$ for a minuscule weight λ.

Remark 14.18: Note that this result holds whether or not Ω^A_M is compatible with the weight grading (by $L_{\mathfrak{h}^\perp_*}(0)$-eigenspaces) of Ω^A_* (cf. Remark 13.13).

Remark 14.19: As in Remark 13.15, Ω^A_* is not in general a module for $\Omega^A_{L(l,0)}$ viewed as a generalized vertex operator algebra because the $L_{\mathcal{G}_l,\mathcal{H}_l}(0)$-eigenspaces of Ω^A_* are not in general finite-dimensional. However, Ω^A_* is a module for $\Omega^A_{L(l,0)}$ viewed simply as a generalized vertex algebra (where the grading restrictions are ignored).

We have not asserted the simplicity (i.e., irreducibility of the algebra as a module for itself) of the algebra $\Omega^A_{L(l,0)}$ in Theorem 14.16 or the irreducibility of the modules Ω^A_M in Proposition 14.17. But in the case $A = B$ (recall (14.6) and (14.10)) – the case in which A is as large as the group H in the definition (14.14) of the group G, we do in fact have:

Theorem 14.20 *The generalized vertex operator algebra $\Omega^B_{L(l,0)}$ is simple and the $\Omega^B_{L(l,0)}$-module Ω^B_M is irreducible for any irreducible $\hat{\mathcal{G}}$-submodule M of V_L. In particular, equipped with the $\mathbf{Q}$-grading defined by the $L^B_{\mathcal{G}_l,\mathcal{H}_l}(0)$-eigenspace decomposition, Ω^B_* is a (usually infinite) direct sum of irreducible $\Omega^B_{L(l,0)}$-modules.*

Proof Write $\Omega_M = M \cap \Omega_*$, as in (14.83). Let ξ be the covering map from Ω_M to Ω^B_M :

$$(14.90)\quad \begin{aligned} \xi : \Omega_M &\rightarrow \Omega^B_M \\ v &\mapsto v + W_M \cap \Omega_* \end{aligned}$$

(recall (14.35) and (14.36)). Let U be a nonzero $\Omega^B_{L(l,0)}$-submodule of Ω^B_M. It is enough to prove that $\xi^{-1}(U) = \Omega_M$. Since by hypothesis U is G-graded, i.e.,

(14.91)
$$U = \coprod_{g \in G} U^g,$$

we see that

(14.92)
$$\xi^{-1}(U) = \sum_{g \in G} \xi^{-1}(U^g).$$

First we prove that $\xi^{-1}(U)$ is stable under the component operators of $Y_*(i(x_\alpha), z)$ for $\alpha \in \Delta$. Let $u \in \xi^{-1}(U)$. Then $\xi(u) = u + W_M \cap \Omega_* \in U$ and $v_n u + W_M \cap \Omega_* = v_n(u + W_M \cap \Omega_*) \in U$ where $v = i(x_\alpha)$ and $n \in \mathbf{Q}$. Thus $v_n u \in \xi^{-1}(U)$. Next we show that $\xi^{-1}(U)$ contains a common eigenvector for all the operators $h(0)$ for $h \in \mathcal{H}$. Choose $g \in G$ such that $U^g \neq 0$ and let $\mu \in P$ be such that $\frac{1}{l}\mu_* + 2Q_* = g$ (recall (14.14)). Then by (14.83)-(14.85) we may take some

$$u = \sum_{\beta \in 2Q} u^{\frac{1}{l}\mu_* + \beta_*} \in \xi^{-1}(U^g)$$

(a finite sum) such that $\xi(u) \neq 0$, where

$$u^{\frac{1}{l}\mu_* + \beta_*} \in \Omega_M^{\frac{1}{l}\mu_* + \beta_*}.$$

Consider the vector

$$x = \sum_{\beta \in 2Q} e_{-\beta_*} \cdot u^{\frac{1}{l}\mu_* + \beta_*} \in \Omega_M^{\frac{1}{l}\mu_*}.$$

Since $e_{-\beta_*} \in B$ (it is here that we use the hypothesis $A = B$), we find that $\xi(x) = \xi(u) \in U$. Thus $x \in \xi^{-1}(U)$ is such a common eigenvector. By the proof of (14.44) we now see that

$$\Omega_M = \mathrm{span}\left\{ v_{n_1}^{(1)} \cdots v_{n_j}^{(j)} \cdot x \,\middle|\, \text{each } v^{(i)} \in \{i(x_\alpha) | \alpha \in \Delta\} \cup \{\iota(1)\} \right\} \subset \xi^{-1}(U),$$

as desired. □

Remark 14.21: Note that the assumption $A = B$, which allowed us to prove irreducibility, essentially says that the group A used to form our quotient spaces contains enough operators to relate the G-grading "tightly" to the action of $\mathcal{H}$.

Remark 14.22: The operators $z^{1-1/l} Y_*(i(x_\alpha), z)$ and $z^{1-1/l} Y_*^B(i(x_\alpha) + W_{L(l,0)}, z)$ ($\alpha \in \Delta$) were introduced in [LP1], where they were denoted $U(\alpha, \zeta)$ (where $\zeta = z^{-1/l}$, in the current notation), for the affine Lie algebra $A_1^{(1)} \simeq \mathbf{sl}(2,\mathbf{C})\hat{}$. In [LP1], they were also studied in detail in the case $l = 2$, where they were found to satisfy certain Clifford algebra relations, giving a construction of the corresponding standard $A_1^{(1)}$-modules. This construction was also generalized in [LP1] to the standard $B_n^{(1)}$-modules of level 1. In fact, the operators $Y_*(i(x_\alpha), z)$ are slight variants of the operators $Y_*(i(x_\alpha), z)z^{1+\alpha_*/l}$ (cf. (14.45), (14.46)), which are just the Z-operators which were extensively studied for the purpose of constructing bases for standard modules for affine algebras and relating such bases to Rogers-Ramanujan-type partition identities (see [LP1]-[LP3] and [LW2]-[LW5]). From the present point of view, the U-operators are a bit more natural than the Z-operators because the U-operators are essentially vertex operators in the present sense. The quotient vertex operators Y_*^B are even more natural than the operators Y_* since $\Omega_{L(l,0)}^B$ is a generalized vertex operator algebra, and $\Omega_{L(l,0)}^B$ is simple and Ω_M^B is an irreducible $\Omega_{L(l,0)}^B$-module for any irreducible $\hat{\mathcal{G}}$-submodule M of V_L.

It is illuminating to work out all this structure in the interesting and important special case $\mathcal{G} = \mathbf{sl}(2,\mathbf{C})$. We fix a root α, which we view as positive. In this case,

$$Q = \mathbf{Z}\alpha, \quad P = \tfrac{1}{2}\mathbf{Z}\alpha,$$
$$L = \tfrac{1}{2}\mathbf{Z}\alpha_1 \oplus \cdots \oplus \tfrac{1}{2}\mathbf{Z}\alpha_l, \quad L_0 = \mathbf{Z}\alpha_1 \oplus \cdots \oplus \mathbf{Z}\alpha_l \tag{14.93}$$

(see (13.12)-(13.13)); the form $\langle\cdot,\cdot\rangle$ on L is defined by

$$\langle\frac{1}{2}\alpha_i, \frac{1}{2}\alpha_j\rangle = \frac{1}{2}\delta_{ij} \tag{14.94}$$

for $i, j \in \{1, ..., l\}$; $T = 2l$ (see (14.19)); $B = \langle e_{2\alpha_*}\rangle$ (see (14.6)); A is an arbitrary subgroup of B with $A \neq \langle 1\rangle$; and

$$G \simeq \left(\frac{1}{2l}\mathbf{Z}\alpha\right)/2\mathbf{Z}\alpha \simeq \left(\frac{1}{2l}\mathbf{Z}\right)/2\mathbf{Z} \tag{14.95}$$

(see (14.14)) with the symmetric form given by

$$(\frac{i}{2l} + 2\mathbf{Z}, \frac{j}{2l} + 2\mathbf{Z}) = \frac{ij}{2l} + 2\mathbf{Z} \in \left(\frac{1}{2l}\mathbf{Z}\right)/2\mathbf{Z} \tag{14.96}$$

for $i, j \in \mathbf{Z}$ (see (14.20)); note that the group of values of this form is isomorphic to the group G.

Also, the commutator map $c(\cdot,\cdot)$ is trivial on Q and in fact on our rank 1 lattice P, and hence on L. We may and do assume that the 2-cocycle $\epsilon(\cdot,\cdot)$ associated with our section $\beta \mapsto e_\beta$ for $\beta \in L$ is trivial as well. (It is not necessary to take the central subgroup $\langle \kappa_q \rangle$ of $\hat{L}$ to be the trivial group.)

For any subset $I \subset \{1,...,l\}$, we shall write

$$\alpha_I = \sum_{i \in I} \alpha_i. \tag{14.97}$$

Note that the elements $\frac{1}{2}\alpha_I$ form a set of coset representatives for the 2^l cosets of L_0 in L. We may take the following spaces (for example) as copies of all the $l+1$ inequivalent standard modules of level l for $\hat{\mathcal{G}}$:

$$L(l,i) = L(l, \frac{i}{2}\alpha) = U(\hat{\mathcal{G}}) \cdot v_i \tag{14.98}$$

for $i = 0,...,l$ (using the notation $L(l,\lambda)$ of Chapter 13), where v_i is the highest weight vector given by

$$v_i = \sum_{|I|=i} \iota(e_{\frac{1}{2}\alpha_I}). \tag{14.99}$$

Then we have subspaces $\Omega^A_{L(l,i)}$ of Ω^A_* for $i = 0,...,l$. Note that $d = \dim \mathcal{G} = 3$, the Coxeter number h is 2 and the rank r (of $\mathcal{G}$) is 1 in this case. By Theorem 14.16, Proposition 14.17 and Theorem 14.20, we have:

Theorem 14.23 *The structure*

$$(\Omega^A_{L(l,0)}, Y^A_*, \iota(A), \omega_{\mathcal{G}_l,\mathcal{H}_l} \otimes \iota(A), T, G, (\cdot,\cdot))$$

is a generalized vertex operator algebra with rank equal to $\frac{2(l-1)}{l+2}$ *and* $T = 2l$, *generated by the* $L^A_{\mathcal{G}_l,\mathcal{H}_l}(0)$*-eigenspace of* $\Omega^A_{L(l,0)}$ *of eigenvalue* $1-\frac{1}{l}$, *and* $\Omega^A_{L(l,i)}$ *is an* $\Omega^A_{L(l,0)}$*-module with lowest weight* $\frac{i(l-i)}{2l(l+2)}$, *for* $i = 0,...,l$. *Moreover,* $\Omega^B_{L(l,0)}$ *is a simple generalized vertex operator algebra and the modules* $\Omega^B_{L(l,i)}$ *are irreducible.* □

In order to illustrate the general theory we would like to write down some concrete formulas in the case $\mathbf{sl}(2,\mathbf{C})$. In this case we have a new option for the groups G and A which is not available in general, because now the alternating form $c(\cdot,\cdot)$ is trivial and the group $\hat{L}$ is abelian. We shall modify the structure in the spirit of Remarks 6.15 and 14.4, using the more general notion of module, based on a G-set S. Aside from the illustrating flexibility of our axiomatic setup, this variation will also serve to explain the context of the formulas in [ZF1].

We shall apply Theorem 6.7 to the pair of lattices L_0, L_0. In particular, $T = l$ (recall (14.24)) and G is the l-element cyclic group

$$(14.100)\quad G = \left(\frac{1}{l}Q_*\right)/Q_* \simeq \left(\frac{1}{l}Q\right)/Q \simeq \left(\frac{1}{l}\mathbf{Z}\right)/\mathbf{Z} \simeq \mathbf{Z}/l\mathbf{Z}$$

(cf. (14.14)), equipped with the natural $(\frac{1}{l}\mathbf{Z})/2\mathbf{Z}$-valued bilinear form given by (5.45); in fact, the values of this form actually lie in $(\frac{2}{l}\mathbf{Z})/2\mathbf{Z}(\simeq \mathbf{Z}/l\mathbf{Z})$. We shall also use the setup of Chapters 4-6 for the pair L, L_0, but in a different way from before. Define

$$(14.101)\quad S = \left(\frac{1}{l}P_*\right)/Q_* \simeq \left(\frac{1}{l}P\right)/Q = \left(\frac{1}{2l}Q\right)/Q \simeq \left(\frac{1}{2l}\mathbf{Z}\right)/\mathbf{Z} \simeq \mathbf{Z}/2l\mathbf{Z}.$$

Then S is an additive group and G is a subgroup of S of index 2. In particular, S is a G-set. We have a well-defined $\left(\frac{1}{l}\mathbf{Z}\right)/\mathbf{Z}$-valued function $(\cdot,\cdot)$ on $G \times S$ given by

$$(14.102)\quad \left(\frac{1}{l}\alpha_* + Q_*, \frac{1}{l}\beta_* + Q_*\right) = \frac{1}{l}\langle\alpha,\beta\rangle + \mathbf{Z} \in \left(\frac{1}{l}\mathbf{Z}\right)/\mathbf{Z}$$

for $\alpha \in Q$ and $\beta \in P$. Note that the values of the function $(\cdot,\cdot)$ lie in $(\frac{1}{l}\mathbf{Z})/\mathbf{Z}$ rather than $(\frac{1}{l}\mathbf{Z})/2\mathbf{Z}$, but this is adequate for defining a module (see Remark 6.15). This function satisfies the biadditivity condition (6.64) mod $\mathbf{Z}$. (Note that (14.102) would not give a well-defined function on $S \times S$.) Identifying S with $(\frac{1}{2l}\mathbf{Z})/\mathbf{Z}$ as in (14.101) and viewing G as a subgroup of $(\frac{1}{2l}\mathbf{Z})/\mathbf{Z}$, we see that the form (14.102) becomes: $(\frac{i}{2l} + \mathbf{Z}, \frac{j}{2l} + \mathbf{Z}) = \frac{ij}{2l} + \mathbf{Z}$ for $i \in 2\mathbf{Z}$, $j \in \mathbf{Z}$ (cf. (14.96)).

Recall the notation Ω_* and $\Omega_*(i)$ for $i \in L/L_0$ from (3.29) and (5.57). Let A be a subgroup of the (abelian) group $\langle e_{\alpha_*}\rangle$ with $A \neq 1$. Clearly, the action of A preserves each $\Omega_*(i)$ (cf. (3.12)). As in (5.59) and (5.60) we consider the subspace

$$(14.103)\quad \Omega_*^A = (\Omega_*)_{L/\bar{A}}$$

of $V_{L/\bar{A}}$ (recall (4.23), (4.27) and (4.30)) and we also write

$$(14.104)\quad \Omega_*^A(i) = \Omega_*(i)/(\Omega_*(i) \cap W_{V_L}) \subset \Omega_*^A$$

for $i \in L/L_0$ (recall (4.24)). Then Ω_*^A is naturally $\mathbf{Q}$-graded as in Chapter 4. We also define an S-gradation on $V_{L/\bar{A}}$ and hence on Ω_*^A by

$$(14.105)\qquad (V_{L/\bar{A}})^s = \sum_{b\in\hat{L},\ \bar{b}''+Q_*=s} (M(1)\otimes\iota(b)+W_{V_L})$$

for $s \in S$ (cf. (5.46)). Then the appropriate analogues of (5.48) and (5.52) hold, using (14.102) and (5.45), respectively. An argument analogous to that for Theorem 14.2, using Proposition 5.6 (cf. Remark 14.3), shows:

Theorem 14.24 *The structure*

$$(\Omega_*^A(0), Y_*^A, \iota(A), \omega\otimes\iota(A), l, G, (\cdot,\cdot))$$

is a generalized vertex operator algebra with rank equal to $l-1$, *and* $(\Omega_*^A, Y_*^A, S, (\cdot,\cdot))$ *and* $(\Omega_*^A(i), Y_*^A, S, (\cdot,\cdot))$ *for* $i \in L/L_0$ *are* $\Omega_*^A(0)$-*modules in the sense of Remark 6.15. If* $A = \langle e_{\alpha_*}\rangle$, *the* $\Omega_*^A(0)$-*module* $\Omega_*^A(i)$ *is irreducible for* $i \in L/L_0$.

Now we imitate the proof of Theorem 14.23 in order to obtain an analogue of it. The proof of Theorem 14.16 clearly carries over. Let M be an irreducible $\hat{\mathcal{G}}$-submodule of V_L. Defining W_M and Ω_M^A as in (14.33) and (14.35), we have that Ω_M^A is S-graded, with

$$(14.106)\qquad (\Omega_M^A)^s = \sum_{\mu\in\frac{1}{l}P,\ \mu_*+Q_*=s} (\Omega_M^{\mu_*}+W_M)/W_M \subset (\Omega_*^A)^s$$

(cf. (14.85)). In fact, if $M \simeq L(l, \frac{i}{2}\alpha)$ for $0 \le i \le l$, it is easy to check that

$$(14.107)\qquad \Omega_M^A = \coprod_{g\in G} (\Omega_M^A)^{(\frac{i}{2l}+\mathbf{Z})+g},$$

using (14.82) and the identifications (14.101). (The image in Ω_M^A of the highest weight space of M has S-degree $\frac{i}{2l}+\mathbf{Z}$.) We now see that Proposition 14.17 and Theorem 14.20 also carry over, and we have:

Theorem 14.25 *The structure*

$$(\Omega_{L(l,0)}^A,\ Y_*^A,\ \iota(A),\ \omega_{\mathcal{G}_l,\mathcal{H}_l}\otimes\iota(A),\ T=l, G=(\frac{1}{l}\mathbf{Z})/\mathbf{Z}, (\cdot,\cdot))$$

is a generalized vertex operator algebra with rank equal to $\frac{2(l-1)}{l+2}$, *generated by the* $L_{\mathcal{G}_l,\mathcal{H}_l}^A(0)$-*eigenspace of* $\Omega_{L(l,0)}^A$ *of eigenvalue* $1-\frac{1}{l}$, *and for* $i = 0, ..., l$,

$$(\Omega_{L(l,i)}^A, Y_*^A, (\frac{i}{2l}+\mathbf{Z})+G, (\cdot,\cdot))$$

is an $\Omega^A_{L(l,0)}$-module in the sense of Remark 6.15, equipped with the **Q**-*grading defined by the $L^A_{\mathcal{G}_l,\mathcal{H}_l}(0)$-eigenspace decomposition, with lowest weight $\frac{i(l-i)}{2l(l+2)}$ for $i = 0, ..., l$, where $(\frac{i}{2l} + \mathbf{Z}) + G \subset S$ is the coset of G in S associated with $\frac{i}{2l} + \mathbf{Z}$. If $A = \langle e_{\alpha_*} \rangle$, $\Omega^A_{L(l,0)}$ is a simple generalized vertex operator algebra and each $\Omega^A_{L(l,i)}$ is irreducible.* □

Remark 14.26: For the case $A = \langle e_{\alpha_*} \rangle$, the generalized vertex operator algebra $\Omega^A_{L(l,0)}$ and its irreducible modules $\Omega^A_{L(l,i)}$ $(i = 0, ..., l)$ provide a mathematical framework for the parafermion conformal field theory in [ZF1]. The parafermion operators are exactly the vertex operators associated with certain generators of the generalized vertex operator algebra $\Omega^A_{L(l,0)}$, as we explain below.

For convenience, we abbreviate $(\Omega^A_*)^{m+\mathbf{Z}}$ by $(\Omega^A_*)^m$ for $m \in \frac{1}{2l}\mathbf{Z}$, using the identification (14.101) for S, and we also use this notation for any subspace of Ω^A_*. For $p = 0, ..., l$, set

$$
\begin{aligned}
u_p &= \binom{l}{p}^{-1/2} \sum_{|I|=p} \iota(Ae_{\alpha_I}) \in (\Omega^A_{L(l,0)})^{p/l} \\
u_p^+ &= \binom{l}{p}^{-1/2} \sum_{|I|=p} \iota(Ae_{-\alpha_I}) \in (\Omega^A_{L(l,0)})^{-p/l}
\end{aligned}
\tag{14.108}
$$

(cf. (14.99)). Note that

$$u_0 = u_0^+ = \iota(A), \quad u_l = \iota(Ae_{\alpha_*}), \quad u_l^+ = \iota(Ae_{-\alpha_*}) \tag{14.109}$$

and that

$$u_1 = l^{-1/2} \sum_{i=1}^{l} \iota(Ae_{\alpha_i}), \quad u_1^+ = l^{-1/2} \sum_{i=1}^{l} \iota(Ae_{-\alpha_i}). \tag{14.110}$$

We shall study the vertex operators $Y^A_*(u_p, z)$ and $Y^A_*(u_p^+, z)$ in detail, initially with A arbitrary, and below, with $A = \langle e_{\alpha_*} \rangle$. We shall regard operators such as e_{α_*} as acting on Ω^A_*. First we have:

Proposition 14.27 *(a) The operators $Y^A_*(u_0, z)$ and $Y^A_*(u_0^+, z)$ are the identity operator:*

$$Y^A_*(u_0, z) = Y^A_*(u_0^+, z) = 1. \tag{14.111}$$

(b) For $0 \leq p \leq l$,

(14.112) $$u_p = e_{\alpha_*} \cdot u_{l-p}^+$$

or equivalently,

(14.113) $$Y_*^A(u_p, z) = e_{\alpha_*} Y_*^A(u_{l-p}^+, z)$$

(see (3.47), (4.42)). In particular (cf. (14.32)),

(14.114) $$Y_*^A(u_l, z) = e_{\alpha_*}, \quad Y_*^A(u_l^+, z) = e_{-\alpha_*}.$$

(c) We have

(14.115) $$\begin{aligned} Y_*^A(u_1, z) &= l^{-1/2} Y_*^A(i(x_\alpha) + W_{L(l,0)}, z) \\ Y_*^A(u_1^+, z) &= l^{-1/2} Y_*^A(i(x_{-\alpha}) + W_{L(l,0)}, z) \end{aligned}$$

(recall (13.16) and Remark 14.7).
(d) The weights of the elements u_p and u_p^+ are as follows:

(14.116) $$\mathrm{wt}\, u_p = \mathrm{wt}\, u_p^+ = \frac{p(l-p)}{l}$$

(cf. Proposition 14.15 and (4.15)). □

A straightforward verification using the definitions (3.38) and (4.39), formula (3.51) and Proposition 14.27 shows us that the operators $Y^*(u_p, z)$ can be built up from "shorter" ones, in particular, from $Y_*^A(u_1, z)$:

Proposition 14.28 *For $0 \le p, q \le l$,*

(14.117) $$\begin{aligned} &\lim_{z_i \to z} (z_1 - z_2)^{2pq/l} Y_*^A(u_p, z_1) Y_*^A(u_q, z_2) \\ &= \begin{cases} c_{pq} Y_*^A(u_{p+q}, z) & \text{if} \quad p+q \le l \\ 0 & \text{if} \quad p+q > l, \end{cases} \end{aligned}$$

where

(14.118) $$c_{pq} = \left[\frac{(l-p)!(l-q)!(p+q)!}{p!q!(l-p-q)!l!} \right]^{\frac{1}{2}}.$$

In particular, for $p + q = l$,

(14.119) $$\lim_{z_i \to z} (z_1 - z_2)^{2pq/l} Y_*^A(u_p, z_1) Y_*^A(u_q, z_2) = e_{\alpha_*}.$$ □

The following generalized Jacobi identity is an immediate consequence of (5.52):

Proposition 14.29 *Let* $i, j \in 2\mathbf{Z}$, $k \in \mathbf{Z}$ *and let* $u \in (\Omega_*^A(0))^{i/2l}$, $v \in (\Omega_*^A(0))^{j/2l}$, $w \in (\Omega_*^A)^{k/2l}$. *Then*

$$z_0^{-1}\left(\frac{z_1 - z_2}{z_0}\right)^{ij/2l} \delta\left(\frac{z_1 - z_2}{z_0}\right) Y_*^A(u, z_1) Y_*^A(v, z_2) w$$

$$(14.120) \quad - z_0^{-1} \delta\left(\frac{z_2 - z_1}{z_0}\right)^{ij/2l} \left(\frac{z_2 - z_1}{-z_0}\right) Y_*^A(v, z_2) Y_*^A(u, z_1) w$$

$$= z_2^{-1}\left(\frac{z_1 - z_0}{z_2}\right)^{-ik/2l} \delta\left(\frac{z_1 - z_0}{z_2}\right) Y_*^A(Y_*^A(u, z_0)v, z_2) w. \quad \square$$

Applying the generalized Jacobi identity (14.120) to $u_p \in (\Omega_{L(l,0)}^A)^{p/l}$ and $u_q \in (\Omega_{L(l,0)}^A)^{q/l}$, and equating the coefficients of four powers of z_0^0 (restricting our attention to $p + q = l$ in the last two cases), we get:

Proposition 14.30 *For* $0 \le p, q \le l$ *and* $k \in \mathbf{Z}$, *as operators on* $(\Omega_*^A)^{k/2l}$,

$$(14.121) \quad \begin{aligned} &(z_1 - z_2)^{2pq/l} Y_*^A(u_p, z_1) Y_*^A(u_q, z_2) \\ &-(z_2 - z_1)^{2pq/l} Y_*^A(u_q, z_2) Y_*^A(u_p, z_1) = 0 \end{aligned}$$

$$(14.122) \quad \begin{aligned} &(z_1 - z_2)^{2pq/l-1} Y_*^A(u_p, z_1) Y_*^A(u_q, z_2) \\ &+ (z_2 - z_1)^{2pq/l-1} Y_*^A(u_q, z_2) Y_*^A(u_p, z_1) \\ &= \begin{cases} c_{pq} Y_*^A(u_{p+q}, z_2) z_2^{-1} \left(\frac{z_1}{z_2}\right)^{-pk/l} \delta\left(\frac{z_1}{z_2}\right) & \text{if } p + q \le l \\ 0 & \text{if } p + q > l, \end{cases} \end{aligned}$$

and if $p + q = l$,

$$(14.123) \quad \begin{aligned} &(z_1 - z_2)^{2pq/l-2} Y_*^A(u_p, z_1) Y_*^A(u_q, z_2) \\ &-(z_2 - z_1)^{2pq/l-2} Y_*^A(u_q, z_2) Y_*^A(u_p, z_1) \\ &= -e_{\alpha_*} z_2^{-1} \frac{\partial}{\partial z_1}\left(\left(\frac{z_1}{z_2}\right)^{-pk/l} \delta\left(\frac{z_1}{z_2}\right)\right) \end{aligned}$$

$$
\begin{aligned}
&(z_1 - z_2)^{2pq/l-3} Y_*^A(u_p, z_1) Y_*^A(u_q, z_2) \\
&+(z_2 - z_1)^{2pq/l-3} Y_*^A(u_q, z_2) Y_*^A(u_p, z_1) \\
&= \frac{pq(l+2)}{l(l-1)} e_{\alpha_*} L^A_{\mathcal{G}_l,\mathcal{H}_l}(z_2) z_2^{-1} \left(\frac{z_1}{z_2}\right)^{-pk/l} \delta\left(\frac{z_1}{z_2}\right) \\
&+\frac{1}{2} e_{\alpha_*} z_2^{-1} \frac{\partial^2}{\partial z_1^2} \left(\left(\frac{z_1}{z_2}\right)^{-pk/l} \delta\left(\frac{z_1}{z_2}\right)\right). \quad \square
\end{aligned}
\tag{14.124}
$$

From now on we assume that $A = \langle e_{\alpha_*} \rangle$ and we write

$$\psi_p(z) = Y_*^A(u_p, z), \quad \psi_p^+(z) = Y_*^A(u_p^+, z) \tag{14.125}$$

$$T(z) = L^A_{\mathcal{G}_l,\mathcal{H}_l}(z). \tag{14.126}$$

Then e_{α_*} acts as the identity operator on Ω_*^A, and we have:

Proposition 14.31 *(a) For* $0 \leq p \leq l$,

$$\psi_p(z) = \psi_{l-p}^+(z) \tag{14.127}$$

and

$$\psi_0(z) = \psi_0^+(z) = \psi_l(z) = \psi_l^+(z) = 1. \tag{14.128}$$

(b) On $(\Omega_*^A)^{k/2l}$ $(k \in \mathbf{Z})$, $\psi_p(z)$ *has an expansion of the form*

$$\psi_p(z)|_{(\Omega_*^A)^{k/2l}} = \sum_{n \in \frac{pk}{l} + \mathbf{Z}} (u_p)_n z^{-n-1} \tag{14.129}$$

and

$$(u_p)_n (\Omega_*^A)^{k/2l} \subset (\Omega_*^A)^{(k+2p)/2l}. \quad \square \tag{14.130}$$

In Propositions 14.28 and 14.30, replacing $Y_*^A(u_p, z)$, $Y_*^A(u_p^+, z)$, e_{α_*} and $L^A_{\mathcal{G}_l,\mathcal{H}_l}(z)$ by $\psi_p(z)$, $\psi_p^+(z)$, 1 and $T(z)$, respectively, we get:

Proposition 14.32 *For* $0 \leq p, q \leq l$,

$$\lim_{z_i \to z} (z_1 - z_2)^{2pq/l} \psi_p(z_1) \psi_q(z_2) = \begin{cases} c_{pq} \psi_{p+q}(z) & \text{if } p+q \leq l \\ 0 & \text{if } p+q > l. \end{cases} \tag{14.131}$$

In particular, for $p + q = l$,

$$\lim_{z_i \to z} (z_1 - z_2)^{2pq/l} \psi_p(z_1) \psi_q(z_2) = 1. \quad \square \tag{14.132}$$

Theorem 14.33 *For $0 \le p, q \le l$, as operators on $(\Omega_*^A)^{k/2l}$ $(k \in \mathbf{Z})$,*

$$(z_1 - z_2)^{2pq/l}\psi_p(z_1)\psi_q(z_2) - (z_2 - z_1)^{2pq/l}\psi_q(z_2)\psi_p(z_1) = 0 \tag{14.133}$$

$$(z_1 - z_2)^{2pq/l-1}\psi_p(z_1)\psi_q(z_2) + (z_2 - z_1)^{2pq/l-1}\psi_q(z_2)\psi_p(z_1)$$
$$= \begin{cases} c_{pq}\psi_{p+q}(z_2)z_2^{-1}\left(\dfrac{z_1}{z_2}\right)^{-pk/l}\delta\left(\dfrac{z_1}{z_2}\right) & \text{if } p+q \le l \\ 0 & \text{if } p+q > l, \end{cases} \tag{14.134}$$

and if $p + q = l$,

$$(z_1 - z_2)^{2pq/l-2}\psi_p(z_1)\psi_q(z_2) - (z_2 - z_1)^{2pq/l-2}\psi_q(z_2)\psi_p(z_1)$$
$$= -z_2^{-1}\frac{\partial}{\partial z_1}\left(\left(\frac{z_1}{z_2}\right)^{-pk/l}\delta\left(\frac{z_1}{z_2}\right)\right) \tag{14.135}$$

$$(z_1 - z_2)^{2pq/l-3}\psi_p(z_1)\psi_q(z_2) + (z_2 - z_1)^{2pq/l-3}\psi_q(z_2)\psi_p(z_1)$$
$$= \frac{pq(l+2)}{l(l-1)}T(z_2)z_2^{-1}\left(\frac{z_1}{z_2}\right)^{-pk/l}\delta\left(\frac{z_1}{z_2}\right)$$
$$+\frac{1}{2}z_2^{-1}\frac{\partial^2}{\partial z_1^2}\left(\left(\frac{z_1}{z_2}\right)^{-pk/l}\delta\left(\frac{z_1}{z_2}\right)\right). \quad \square \tag{14.136}$$

Remark 14.34: The operators ψ and ψ^+ correspond exactly to the parafermion operators in [ZF1]. These operators were studied in [ZF1] from the point of view of conformal field theory. The group G corresponds to the cyclic symmetry group in [ZF1]. Proposition 14.32 and Theorem 14.33 correspond to formulas (3.7a)-(3.7c) of [ZF1] (with (3.7b) suitably corrected). The operators $\psi_1(z)$ and $\psi_1^+(z)$ generate the generalized vertex operator algebra $\Omega_{L(l,0)}^A$ (recall Propositions 14.9 and 14.27(c); see also (14.131)), and these operators are essentially the generating Z-operators or U-operators of [LP1] (see Remark 14.22). Equating coefficients of the monomials in z_1 and z_2 in the identities in Theorem 14.33 gives identities involving infinite sums of products of the component operators $(u_p)_n$ (recall (14.129)). Some of these identities are used in [ZF1]. Such "generalized commutation relations" and "generalized anticommutation relations" were introduced in [LP1]-[LP3] and [LW2]-[LW5] in order to "straighten" monomials in the component operators, for the purpose of finding bases of spaces such as the $\Omega_{L(l,i)}^A$. In fact, this was the problem which motivated the introduction of the theory of Z-algebras.

Bibliography

[A] F. Akman, The semi-infinite Weil complex of a graded Lie algebra, Ph.D. thesis, Yale University, 1993.

[BHN] T. Banks, D. Horn and H. Neuberger, Bosonization of the $SU(N)$ Thirring models, *Nucl. Phys.* **B108** (1976), 119.

[BH] K. Bardakci and M. B. Halpern, New dual quark models, *Phys. Rev.* **D3** (1971), 2493-2506.

[BPZ] A. A. Belavin, A. M. Polyakov and A. B. Zamolodchikov, Infinite conformal symmetries in two-dimensional quantum field theory, *Nucl. Phys.* **B241** (1984), 333-380.

[Bor] R. E. Borcherds, Vertex algebras, Kac-Moody algebras, and the Monster, *Proc. Natl. Acad. Sci. USA* **83** (1986), 3068-3071.

[Bou] N. Bourbaki, Groupes et algèbres de Lie, Chaps. 4, 5, 6, Hermann, Paris, 1968.

[Br] K. S. Brown, Cohomology of Groups, *Graduate Texts in Mathematics* **87**, Springer-Verlag, New York, 1982.

[C] S. Capparelli, On some representations of twisted affine Lie algebras and combinatorial identities, *J. Algebra* **154** (1993), 335-355.

[DL1] C. Dong and J. Lepowsky, A Jacobi identity for relative vertex operators and the equivalence of Z-algebras and parafermion algebras, in: *Proc. XVIIth Intl. Colloq. on Group Theoretical Methods in Physics,* Ste-Adèle, June, 1988, ed. Y. Saint-Aubin and L. Vinet, World Scientific, Singapore, 1989, 235-238.

[DL2] C. Dong and J. Lepowsky, Abelian intertwining algebras – a generalization of vertex operator algebras, in: *Algebraic Groups and Generalizations, Proc. 1991 American Math. Soc. Summer Research Institute,* ed. by W. Haboush and B. Parshall, *Proc. Symp. Pure Math.,* American Math. Soc., 1993.

[DL3] C. Dong and J. Lepowsky, The algebraic structure of relative twisted vertex operators, to appear.

[E] S. Eilenberg, Homotopy groups and algebraic homology theories, *Proc. Intl. Congress of Mathematicians,* Vol. I, 1950, 350-353.

[EML1] S. Eilenberg and S. Mac Lane, On the groups $H(\pi, n)$, I, *Annals of Math.* **58** (1953), 55-106.

[EML2] S. Eilenberg and S. Mac Lane, On the groups $H(\pi, n)$, II, *Annals of Math.* **70** (1954), 49-137.

[FFR] A. J. Feingold, I. B. Frenkel and J. F. X. Ries, Spinor construction of vertex operator algebras, triality and $E_8^{(1)}$, *Contemporary Math.* **121,** 1991.

[FL] A. J. Feingold and J. Lepowsky, The Weyl-Kac character formula and power series identities, *Advances in Math.* **29** (1978), 271-309.

[FradK] E. Fradkin and L. P. Kadanoff, Disorder variables and parafermions in two-dimensional statistical mechanics, *Nucl. Phys.* **B170** (1980), 1-15.

[FHL] I. B. Frenkel, Y.-Z. Huang and J. Lepowsky, On axiomatic approaches to vertex operator algebras and modules, preprint, 1989; *Memoirs American Math. Soc.* **104**, 1993.

[FK] I. B. Frenkel and V. G. Kac, Basic representations of affine Lie algebras and dual resonance models, *Invent. Math.* **62** (1980), 23-66.

[FLM1] I. B. Frenkel, J. Lepowsky and A. Meurman, A natural representation of the Fischer-Griess Monster with the modular function J as character, *Proc. Natl. Acad. Sci. USA* **81** (1984), 3256-3260.

[FLM2] I. B. Frenkel, J. Lepowsky and A. Meurman, Vertex operator calculus, in: *Mathematical Aspects of String Theory, Proc.*

1986 Conference, San Diego, ed. by S.-T. Yau, World Scientific, Singapore, 1987, 150-188.

[FLM3] I. B. Frenkel, J. Lepowsky and A. Meurman, Vertex Operator Algebras and the Monster, *Pure and Applied Math.,* Vol. 134, Academic Press, 1988.

[FZ] I. B. Frenkel and Y. Zhu, Vertex operator algebras associated to representations of affine and Virasoro algebras, *Duke Math. J.* **66** (1992), 123-168.

[Gep] D. Gepner, New conformal field theories associated with Lie algebras and their partition functions, *Nucl. Phys.* **B290** (1987), 10-24.

[GW] D. Gepner and E. Witten, String theory on group manifolds, *Nucl. Phys.* **B278** (1986), 493-549.

[Ger] M. Gerstenhaber, On the deformation of rings and algebras: II, *Ann. of Math.* **84** (1966), 1-19.

[Go] P. Goddard, Meromorphic conformal field theory, in: *Infinite Dimensional Lie Algebras and Groups, Advanced Series in Math. Physics,* Vol. 7, ed. by V. Kac, World Scientific, Singapore, 1989, 556.

[GKO] P. Goddard, A. Kent and D. Olive, Unitary representations of the Virasoro and super-Virasoro algebras, *Comm. Math. Phys.* **103** (1986), 105-119.

[Gu] H. Guo, On abelian intertwining algebras and modules, Ph.D. thesis, Rutgers University, 1993.

[Ha] M. B. Halpern, Quantum "solitons" which are $SU(N)$ fermions, *Phys. Rev.* **D12** (1975), 1684-1699.

[Ho] D. F. Holt, An interpretation of the cohomology groups $H^n(G, M)$, *J. Algebra* **60** (1979), 307-320.

[Hu] C. Husu, Extensions of the Jacobi identity for vertex operators, and standard $A_1^{(1)}$-modules, Ph.D. thesis, Rutgers University, 1990, and *Memoirs American Math. Soc.,* to appear.

[JS] A. Joyal and R. Street, Braided monoidal categories, *Macquarie Mathematics Reports,* Macquarie University, Australia, 1986.

[K] V. G. Kac, Infinite dimensional Lie algebras, 3rd ed., Cambridge Univ. Press, Cambridge, 1990.

[KZ] V. G. Knizhnik and A. B. Zamolodchikov, Current algebra and Wess-Zumino model in two dimensions, *Nucl. Phys.* **B247** (1984), 83-103.

[L1] J. Lepowsky, Lectures on Kac-Moody Lie algebras, Univ. Paris VI, 1978.

[L2] J. Lepowsky, Calculus of twisted vertex operators, *Proc. Natl. Acad Sci. USA* **82** (1985), 8295-8299.

[L3] J. Lepowsky, Vertex operator algebras and partition identities, to appear.

[LP1] J. Lepowsky and M. Primc, Standard modules for type one affine Lie algebras, in: Number Theory, New York, 1982, *Lecture Notes in Math.* **1052**, Springer-Verlag, 1984, 194-251.

[LP2] J. Lepowsky and M. Primc, Structure of the standard modules for the affine Lie algebra $A_1^{(1)}$, *Contemporary Math.* **46**, 1985.

[LP3] J. Lepowsky and M. Primc, Structure of the standard $A_1^{(1)}$-modules in the homogeneous picture, in: *Vertex Operators in Mathematics and Physics, Proc. 1983 M.S.R.I. Conference,* ed. by J. Lepowsky, S. Mandelstam and I. M. Singer, Publ. Math. Sciences Res. Inst. #3, Springer-Verlag, New York, 1985, 143-162.

[LW1] J. Lepowsky and R.L. Wilson, Construction of the affine Lie algebra $A_1^{(1)}$, *Comm. Math. Phys.* **62** (1978), 43-53.

[LW2] J. Lepowsky and R.L. Wilson, A new family of algebras underlying the Rogers-Ramanujan identities and generalizations, *Proc. Natl. Acad. Sci. USA* **78** (1981), 7245-7248.

[LW3] J. Lepowsky and R.L. Wilson, The structure of standard modules, I: Universal algebras and the Rogers-Ramanujan identities, *Invent. Math.* **77** (1984), 199-290.

[LW4] J. Lepowsky and R.L. Wilson, The structure of standard modules, II: The case $A_1^{(1)}$, principal gradation, *Invent. Math.* **79** (1985), 417-442.

[LW5] J. Lepowsky and R.L. Wilson, Z-algebras and the Rogers-Ramanujan identities, in: *Vertex Operators in Mathematics and Physics, Proc. 1983 M.S.R.I. Conference,* ed. by J. Lepowsky, S. Mandelstam and I. M. Singer, Publ. Math. Sciences Res. Inst. #3, Springer-Verlag, New York, 1985, 97-142.

[LZ] B. H. Lian and G. J. Zuckerman, New perspectives on the BRST-algebraic structure of string theory, to appear.

[ML] S. Mac Lane, Cohomology theory of abelian groups, *Proc. Intl. Congress of Mathematicians,* Vol. II, 1950, 8-14.

[Ma] M. Mandia, Structure of the level one standard modules for the affine Lie algebras $B_l^{(1)}$, $F_4^{(1)}$ and $G_2^{(1)}$, *Memoirs American Math. Soc.* **362**, 1987.

[MP] A. Meurman and M. Primc, Annihilating ideals of standard modules of $sl(2, \mathbf{C})\tilde{}$ and combinatorial identities, *Advances in Math.* **64** (1987), 177-240.

[Mi1] K. C. Misra, Realization of the level two standard $sl(2k+1, \mathbf{C})\tilde{}$-modules, *Trans. Amer. Math. Soc.* **316** (1989), 295-309.

[Mi2] K. C. Misra, Realization of the level one standard $\tilde{C}_{2k+1}$-modules, *Trans. Amer. Math. Soc.* **321** (1990), 483-504.

[Mi3] K. C. Misra, Level one standard modules for affine symplectic Lie algebras, *Math. Ann.* **287** (1990), 287-302.

[Mi4] K. C. Misra, Level two standard $\tilde{A}_n$-modules, *J. Algebra* **137** (1991), 56-76.

[MS] G. Moore and N. Seiberg, Classical and quantum conformal field theory, *Comm. Math. Phys.* **123** (1989), 177-254.

[Mo] G. Mossberg, A generalized Jacobi identity in axiomatic vertex operator algebras, licentiate thesis, Univ. of Lund, 1992, Report Univ. of Lund, Dept. of Math., 1992:10.

[P1] M. Primc, Standard representations of $A_n^{(1)}$, in: *Infinite Dimensional Lie Algebras and Groups, Advanced Series in Math. Physics,* Vol. 7, ed. by V. Kac, World Scientific, Singapore, 1989, 273-282.

[P2] M. Primc, Vertex operator construction of standard modules for $A_n^{(1)}$, *Pacific J. Math.,* to appear.

[Se] G. Segal, Unitary representations of some infinite-dimensional groups, *Comm. Math. Phys.* **80** (1981), 301-342.

[Su] H. Sugawara, A field theory of currents, *Phys. Rev.* **170** (1968), 1659-1662.

[TK] A. Tsuchiya and Y. Kanie, Vertex operators in conformal field theory on $\mathbf{P}^1$ and monodromy representations of braid group, in: *Conformal Field Theory and Solvable Lattice Models, Advanced Studies in Pure Math.,* Vol. 16, Kinokuniya Company Ltd., Tokyo, 1988, 297-372.

[T] H. Tsukada, Vertex operator superalgebras, *Comm. in Algebra* **18** (1990), 2249-2274.

[V] E. Verlinde, Fusion rules and modular transformations in 2D conformal field theory, *Nucl. Phys.* **B300** (1988), 360-376.

[W1] E. Witten, Non-abelian bosonization in two dimensions, *Comm. Math. Phys.* **92** (1984), 455-472.

[W2] E. Witten, Physics and geometry, *Proc. Intl. Congress of Mathematicians, Berkeley, 1986,* Vol. 1, American Math. Soc., 1987, 267-303.

[X] X. Xu, Twisted modules of colored lattice vertex operator superalgebras, to appear.

[ZF1] A. B. Zamolodchikov and V. A. Fateev, Nonlocal (parafermion) currents in two-dimensional conformal quantum field theory and self-dual critical points in Z_N-symmetric statistical systems, *Sov. Phys. JETP* **62** (1985), 215-225.

[ZF2] A. B. Zamolodchikov and V. A. Fateev, Disorder fields in two-dimensional conformal quantum field theory and $N = 2$ extended supersymmetry, *Sov. Phys. JETP* **63** (1986), 913-919.

List of frequently-used symbols, in order of appearance

Chapter 2

$\mathbf{Z}$, $\mathbf{N}$, $\mathbf{Q}$, $\mathbf{R}$, $\mathbf{C}$
κ_n, $\langle\kappa_n\rangle$, $\langle\cdot\rangle$
L, $\langle\cdot,\cdot\rangle$
ν, q
$c_0(\cdot,\cdot)$
$\hat{L}$, $a \mapsto \bar{a}$
$\hat{\nu}$
$\mathfrak{h}$, $\mathfrak{h}_*$, $\mathfrak{h}_*^{\perp}$
h', h''

Chapter 3

$\hat{\mathfrak{h}}$, c
$\mathrm{wt}\,(x \otimes t^n)$
$\hat{\mathfrak{h}}^+$, $\hat{\mathfrak{h}}^-$, $\hat{\mathfrak{h}}_{\mathbf{Z}}$
$M(1)$, $M(k)$
$U(\cdot)$, $S(\cdot)$
$\mathbf{C}\{L\}$, $\mathbf{C}[\hat{L}]$, $\mathbf{C}[L]$, ω_q
$\iota(a)$
z, z_0, z_1, z_2, ...
z^h
V_L
wt
$c(\cdot,\cdot)$
$\alpha(n)$
Ω_*, V_*
$M_*(1)$
$\alpha(z)$
$\overset{\circ}{\circ}\ \overset{\circ}{\circ}$
$Y_*(a,z)$, $Y_*(v,z)$, v_n
$W\{z\}$

Chapter 4

L_0
A, $\bar{A}$
$\mathbf{C}\{L/\bar{A}\}$, $W_{\mathbf{C}\{L\}}$, $\iota(Ab)$, $\mathrm{wt}\,\iota(Ab)$
$V_{L/\bar{A}}$, W_{V_L}
$(\Omega_*)_{L/\bar{A}}$, $(V_*)_{L/\bar{A}}$
$\hat{M}$, $\mathbf{C}\{M\}$, V_M
V_{L_0}, $V_{L_0/\bar{A}}$, $(\Omega_*)_{L_0/\bar{A}}$
$Y_*^A(v,z)$, v_n

Chapter 5

$\delta(z)$, $\delta(z_1/z_2)$
Res_z
$\mathfrak{h}_{\mathbf{Q}}$
L', L''
T
S°
G, H, $(\cdot,\cdot)$
$(V_{L/\bar{A}})^g$
$z^g\delta(z)$
L_0°
λ_i, Λ_i
$V(i)$, $\Omega_*(i)$
Ω_*^A, $\Omega_*^A(i)$
ω, $L(z)$, $L(n)$

Chapter 6

G, T, $(\cdot,\cdot)$
V, V_n, V^g
$Y(v,z)$, v_n
$\mathbf{1}$
ω, $L(n)$
rank V
$(V, Y, \mathbf{1}, \omega, T, G, (\cdot,\cdot))$
z^g

$(\Omega_*^A(0), Y_*^A, \iota(A), \omega \otimes \iota(A),$
$T, G, (\cdot,\cdot))$
(W, Y)
(Ω_*^A, Y_*^A), $(\Omega_*^A(i), Y_*^A)$
S, $(W, Y, S, (\cdot,\cdot))$
$(V, Y, \mathbf{1}, \omega, \mathbf{Z}/2\mathbf{Z})$

Chapter 7

S, $\iota_{i_1 i_2}$
V', $\langle\cdot,\cdot\rangle$
$\iota_{i_1 \cdots i_n}$

Chapter 8

z^r, $(z_1 - z_2)^r$
H_ϵ
R_σ, U_σ, $U(i, \epsilon)$
X_n, I_1
B_n, b_j
$\zeta_i(t)$
$\tilde{X}_n$
$\mathbf{v}$, $Y_{\mathbf{v}}(z_1, ..., z_n)$
$V(g)$
$W(g)$
$H_n(q)$, T_i

Chapter 9

$(V, Y, \mathbf{1}, \omega, T, G, (\cdot,\cdot), c(\cdot,\cdot))$
(W, Y), $(W, Y, S, (\cdot,\cdot))$
$(\Omega_*, Y_*, \mathbf{1}, \omega, T, G, (\cdot,\cdot), c(\cdot,\cdot))$
$(V, Y, \mathbf{1}, \omega, \mathbf{Z}/2\mathbf{Z})$

Chapter 10

$V_1 \otimes \cdots \otimes V_n$

Chapter 11

$\begin{bmatrix} i \\ j\ k \end{bmatrix}$, $\begin{bmatrix} W_i \\ W_j\ W_k \end{bmatrix}$,
$\mathcal{Y}(w, z)$, w_n
$(\mathcal{Y}, +, (\cdot,\cdot)_{ijk}, c_j(\cdot,\cdot))$
$\mathcal{V}_i^{jk}$, N_{jk}^i

Chapter 12

$e^{i\pi\alpha}$, $c(\cdot, \alpha)$
$\mathcal{Y}_{\lambda_j}(w, z)$
$h(j_1, j_2, j_3)$
$\epsilon(j_1, j_2)$
$G = L/L_0$
$p(\cdot,\cdot)$, $r(\cdot,\cdot)$, $s(\cdot,\cdot)$, T
$\mathcal{X}_{g_1, g_2}(w, z)$
$f(\cdot,\cdot)$, $(df)(\cdot,\cdot,\cdot)$
$F(g_1, g_2, g_3)$
$\Omega(g_1, g_2)$
$B(g_1, g_2, g_3)$
$q(g)$, $b(g_1, g_2)$
$(V, Y, \mathbf{1}, \omega, T, G, F(\cdot,\cdot,\cdot), \Omega(\cdot,\cdot))$
$\hat{b}$
$V_1 \otimes \cdots \otimes V_n$

Chapter 13

$\mathcal{G}$, $\langle\cdot,\cdot\rangle$
$\mathcal{H}$, Δ
Q, P
$\hat{P}$, e_α
x_α
$\epsilon(\alpha, \beta)$
$\hat{\mathcal{G}}$, c
$x(z)$
l
L, L_0
$\mathbf{h}$
$c_0(\cdot,\cdot)$, $c(\cdot,\cdot)$, $\epsilon_0(\cdot,\cdot)$
i, $i(h)$, $i(x_\alpha)$
π, $\pi(c)$, $\pi(x(z))$
$x(n)$
$L(l, 0)$

$d = \dim \mathcal{G}$
$h =$ Coxeter number
Ω
$\omega_{\mathcal{G}_l}$
$Y_V(v,z)$, $Y_W(v,z)$
${}^{\times}_{\times}\ {}^{\times}_{\times}$, $:\ :$
$L_{\mathcal{G}_l}(z)$, $L_{\mathcal{G}_l}(n)$,
$L(l,\lambda)$, $h_{l,\lambda}$
ρ
$\mathbf{Y}(v,z)$

Chapter 14

α_*, Q_*
$\mathbf{h}_*$, $\mathbf{h}_*^\perp$
B, e_{α_*}, A
G, T, $(\cdot,\cdot)$
$E^\pm(h,z)$
M, W_M, M^A, Ω_M^A
$\Omega_{L(l,0)}^A$
$\Omega_{L(l,0)}$
$\omega_{\mathcal{H}_l}$, $L_{\mathcal{H}_l}(z)$, $L_{\mathcal{H}_l}(n)$
$\omega_{\mathcal{G}_l,\mathcal{H}_l}$, $L_{\mathcal{G}_l,\mathcal{H}_l}(z)$, $L_{\mathcal{G}_l,\mathcal{H}_l}(n)$
$\omega_{\mathbf{h}_*^\perp}$, $L_{\mathbf{h}_*^\perp}(z)$, $L_{\mathbf{h}_*^\perp}(n)$
$L_{\mathcal{G}_l,\mathcal{H}_l}^A(z)$, $L_{\mathcal{G}_l,\mathcal{H}_l}^A(n)$
$c_{\mathcal{G}_l,\mathcal{H}_l}$
$\Omega_*^{\mu_*}$, $\Omega_M^{\mu_*}$, $(\Omega_M^A)^g$
$h_{l,\lambda}^*$
Q, P, L, L_0
G
α_I
$L(l,i)$
v_i
G, S
$(V_{L/\bar{A}})^s$
$(\Omega_M^A)^s$
$(\Omega_*^A)^m$
u_p, u_p^+, u_1, u_1^+
$\psi_p(z)$, $\psi_p^+(z)$
$T(z)$

Index

Progress in Mathematics

Edited by:

J. Oesterlé
Département de Mathématiques
Université de Paris VI
4, Place Jussieu
75230 Paris Cedex 05, France

A. Weinstein
Department of Mathematics
University of California
Berkeley, CA 94720
U.S.A.

Progress in Mathematics is a series of books intended for professional mathematicians and scientists, encompassing all areas of pure mathematics. This distinguished series, which began in 1979, includes authored monographs and edited collections of papers on important research developments as well as expositions of particular subject areas.

We encourage preparation of manuscripts in some form of TeX for delivery in camera-ready copy which leads to rapid publication, or in electronic form for interfacing with laser printers or typesetters.

Proposals should be sent directly to the editors or to: Birkhäuser Boston, 675 Massachusetts Avenue, Cambridge, MA 02139, U. S. A.

1 GROSS. Quadratic Forms in Infinite-Dimensional Vector Spaces
2 PHAM. Singularités des Systèmes Différentiels de Gauss-Manin
3 OKONEK/SCHNEIDER/SPINDLER. Vector Bundles on Complex Projective Spaces
4 AUPETIT. Complex Approximation, Proceedings, Quebec, Canada, July 3-8, 1978
5 HELGASON. The Radon Transform
6 LION/VERGNE. The Weil Representation, Maslov Index and Theta Series
7 HIRSCHOWITZ. Vector Bundles and Differential Equations Proceedings. Nice, France, June 12-17, 1979
8 GUCKENHEIMER/MOSER/NEWHOUSE. Dynamical Systems, C.I.M.E. Lectures. Bressanone, Italy, June, 1978
9 SPRINGER. Linear Algebraic Groups
10 KATOK. Ergodic Theory and Dynamical Systems I
11 BALSEV. 18th Scandinavian Conferess of Mathematicians, Aarhus, Denmark, 1980
12 BERTIN. Séminaire de Théorie des Nombres, Paris 1979-80
13 HELGASON. Topics in Harmonic Analysis on Homogeneous Spaces
14 HANO/MARIMOTO/MURAKAMI/OKAMOTO/OZEKI. Manifolds and Lie Groups: Papers in Honor of Yozo Matsushima
15 VOGAN. Representations of Real Reductive Lie Groups
16 GRIFFITHS/MORGAN. Rational Homotopy Theory and Differential Forms
17 VOVSI. Triangular Products of Group Representations and Their Applications
18 FRESNEL/VAN DER PUT. Géométrie Analytique Rigide et Applications
19 ODA. Periods of Hilbert Modular Surfaces
20 STEVENS. Arithmetic on Modular Curves

21 KATOK. Ergodic Theory and Dynamical Systems II
22 BERTIN. Séminaire de Théorie des Nombres, Paris 1980-81
23 WEIL. Adeles and Algebraic Groups
24 LE BARZ/HERVIER. Enumerative Geometry and Classical Algebraic Geometry
25 GRIFFITHS. Exterior Differential Systems and the Calculus of Variations
26 KOBLITZ. Number Theory Related to Fermat's Last Theorem
27 BROCKETT/MILLMAN/SUSSMAN. Differential Geometric Control Theory
28 MUMFORD. Tata Lectures on Theta I
29 FRIEDMAN/MORRISON. Birational Geometry of Degenrations
30 YANO/KON. CR Submanifolds of Kaehlerian and Sasakian Manifolds
31 BERTRAND/WALDSCHMIDT. Approximations Diophantiennes et Nombres Transcendants
32 BOOKS/GRAY/REINHART. Differential Geometry
33 ZUILY. Uniqueness and Non-Uniqueness in the Cauchy Problem
34 KASHIWARA. Systems of Micro-differential Equations
35 ARTIN/TATE. Arithmetic and Geometry: Papers Dedicated to I. R. Shafarevich on the Occasion of His Sixtieth Birthday. Vol. 1
36 ARTIN/TATE. Arithmetic and Geometry: Papers Dedicated to I. R. Shafarevich on the Occasion of His Sixtieth Birthday. Vol. II
37 DE MONVEL. Mathématique et Physique
38 BERTIN. Séminaire de Théorie des Nombres, Paris 1981-82
39 UENO. Classification of Algebraic and Analytic Manifolds
40 TROMBI. Representation Theory of Reductive Groups
41 STANLEY. Combinatorics and Commutative Algebra
42 JOUANOLOU. Théorèmes de Bertini et Applications
43 MUMFORD. Tata Lectures on Theta II
44 KAC. Infitine Dimensional Lie Algebras
45 BISMUT. Large deviations and the Malliavin Calculus
46 SATAKE/MORITA. Automorphic Forms of Several Variables, Taniguchi Symposium, Katata, 1983
47 TATE. Les Conjectures de Stark sur les Fonctions L d'Artin en $s = 0$
48 FRÖLICH. Classgroups and Hermitian Modules
49 SCHLICHTKRULL. Hyperfunctions and Harmonic Analysis on Symmetric Spaces
50 BOREL ET AL. Intersection Cohomology
51 BERTIN/GOLDSTEIN. Séminaire de Théorie des Nombres, Paris 1982-83
52 GASQUI/GOLDSCHMIDT. Déformations Infinitesimales des Structures Conformes Plates
53 LAURENT. Théorie de la Deuxième Microlocalisation dans le Domaine Complexe
54 VERDIER/LE POTIER. Module des Fibres Stables sur les Courbes Algébriques: Notes de l'Ecole Normale Supérieure, Printemps, 1983
55 EICHLER/ZAGIER. The Theory of Jacobi Forms
56 SHIFFMAN /SOMMESE. Vanishing Theorems on Complex Manifolds
57 RIESEL. Prime Numbers and Computer Methods for Factorization
58 HELFFER/NOURRIGAT. Hypoellipticité Maximale pour des Opérateurs Polynomes de Champs de Vecteurs
59 GOLDSTEIN. Séminaire de Théorie des Nombres, Paris 1983–84
60 PROCESI. Geometry Today: Giornate Di Geometria, Roma. 1984

61 BALLMANN/GROMOV/SCHROEDER. Manifolds of Nonpositive Curvature
62 GUILLOU/MARIN. A la Recherche de la Topologie Perdue
63 GOLDSTEIN. Séminaire de Théorie des Nombres, Paris 1984–85
64 MYUNG. Malcev-Admissible Algebras
65 GRUBB. Functional Calculus of Pseudo-Differential Boundary Problems
66 CASSOU-NOGUES/TAYLOR. Elliptic Functions and Rings and Integers
67 HOWE. Discrete Groups in Geometry and Analysis: Papers in Honor of G.D. Mostow on His Sixtieth Birthday
68 ROBERT. Autour de L'Approximation Semi-Classique
69 FARAUT/HARZALLAH. Deux Cours d'Analyse Harmonique
70 ADOLPHSON/CONREY/GHOSH/YAGER. Analytic Number Theory and Diophantine Problems: Proceedings of a Conference at Oklahoma State University
71 GOLDSTEIN. Séminaire de Théorie des Nombres, Paris 1985–86
72 VAISMAN. Symplectic Geometry and Secondary Characteristic Classes
73 MOLINO. Riemannian Foliations
74 HENKIN/LEITERER. Andreotti-Grauert Theory by Integral Formulas
75 GOLDSTEIN. Séminaire de Théorie des Nombres, Paris 1986–87
76 COSSEC/DOLGACHEV. Enriques Surfaces I
77 REYSSAT. Quelques Aspects des Surfaces de Riemann
78 BORHO /BRYLINSKI/MACPHERSON. Nilpotent Orbits, Primitive Ideals, and Characteristic Classes
79 MCKENZIE/VALERIOTE. The Structure of Decidable Locally Finite Varieties
80 KRAFT/PETRIE/SCHWARZ. Topological Methods in Algebraic Transformation Groups
81 GOLDSTEIN. Séminaire de Théorie des Nombres, Paris 1987–88
82 DUFLO/PEDERSEN/VERGNE. The Orbit Method in Representation Theory: Proceedings of a Conference held in Copenhagen, August to September 1988
83 GHYS/DE LA HARPE. Sur les Groupes Hyperboliques d'après Mikhael Gromov
84 ARAKI/KADISON. Mappings of Operator Algebras: Proceedings of the Japan-U.S. Joint Seminar, University of Pennsylvania, Philadelphia, Pennsylvania, 1988
85 BERNDT/DIAMOND/HALBERSTAM/HILDEBRAND. Analytic Number Theory: Proceedings of a Conference in Honor of Paul T. Bateman
86 CARTIER/ILLUSIE/KATZ/LAUMON/MANIN/RIBET. The Grothendieck Festschrift: A Collection of Articles Written in Honor of the 60th Birthday of Alexander Grothendieck. Vol. I
87 CARTIER/ILLUSIE/KATZ/LAUMON/MANIN/RIBET. The Grothendieck Festschrift: A Collection of Articles Written in Honor of the 60th Birthday of Alexander Grothendieck. Volume II
88 CARTIER/ILLUSIE/KATZ/LAUMON/MANIN/RIBET. The Grothendieck Festschrift: A Collection of Articles Written in Honor of the 60th Birthday of Alexander Grothendieck. Volume III
89 VAN DER GEER/OORT / STEENBRINK. Arithmetic Algebraic Geometry
90 SRINIVAS. Algebraic K-Theory
91 GOLDSTEIN. Séminaire de Théorie des Nombres, Paris 1988–89
92 CONNES/DUFLO/JOSEPH/RENTSCHLER. Operator Algebras, Unitary Representations, Enveloping Algebras, and Invariant Theory. A Collection of Articles in Honor of the 65th Birthday of Jacques Dixmier

93 AUDIN. The Topology of Torus Actions on Symplectic Manifolds
94 MORA/TRAVERSO (eds.) Effective Methods in Algebraic Geometry
95 MICHLER/RINGEL (eds.) Representation Theory of Finite Groups and Finite Dimensional Algebras
96 MALGRANGE. Equations Différentielles à Coefficients Polynomiaux
97 MUMFORD/NORI/NORMAN. Tata Lectures on Theta III
98 GODBILLON. Feuilletages, Etudes géométriques
99 DONATO /DUVAL/ELHADAD/TUYNMAN. Symplectic Geometry and Mathematical Physics. A Collection of Articles in Honor of J.-M. Souriau
100 TAYLOR. Pseudodifferential Operators and Nonlinear PDE
101 BARKER/SALLY. Harmonic Analysis on Reductive Groups
102 DAVID. Séminaire de Théorie des Nombres, Paris 1989-90
103 ANGER /PORTENIER. Radon Integrals
104 ADAMS /BARBASCH/VOGAN. The Langlands Classification and Irreducible Characters for Real Reductive Groups
105 TIRAO/WALLACH. New Developments in Lie Theory and Their Applications
106 BUSER. Geometry and Spectra of Compact Riemann Surfaces
107 BRYLINSKI. Loop Spaces, Characteristic Classes and Geometric Quantization
108 DAVID. Séminaire de Théorie des Nombres, Paris 1990-91
109 EYSSETTE/GALLIGO. Computational Algebraic Geometry
110 LUSZTIG. Introduction to Quantum Groups
111 SCHWARZ. Morse Homology
112 DONG/LEPOWSKY. Generalized Vertex Algebras and Relative Vertex Operators